SPECTROSCOPY FOR THE BIOLOGICAL SCIENCES

SPECTROSCOPY FOR THE BIOLOGICAL SCIENCES

GORDON G. HAMMES
Department of Biochemistry
Duke University

WILEY-
INTERSCIENCE

A JOHN WILEY & SONS, INC., PUBLICATION

Published by John Wiley & Sons, Inc., Hoboken, New Jersey
Published simultaneously in Canada

For general information on our other products and services or for technical support, please contact our Customer Care Department within the United States at (800) 762-2974, outside the United States at (317) 572-3993 or fax (317) 572-4002.

Wiley also publishes its books in a variety of electronic formats. Some content that appears in print may not be available in electronic formats. For more information about Wiley products, visit our web site at www.wiley.com.

Library of Congress Cataloging-in-Publication Data:
Hammes, Gordon G., 1934–
　Spectroscopy for the biological sciences / Gordon G. Hammes.
　　p. ; cm.
　Companion v. to: Thermodynamics and kinetics for the biological sciences / Gordon G. Hammes. c2000.
　Includes bibliographical references and index.
　ISBN-13 978-0-471-71344-9 (pbk.)
　ISBN-10 0-471-71344-9 (pbk.)
　1. Biomolecules—Spectra.　2. Spectrum analysis.
　[DNLM: 1. Spectrum Analysis.　2. Crystallography, X-Ray.]　I. Hammes, Gordon G., 1934–　Thermodynamics and kinetics for the biological sciences.　II. Title.
　QP519. 9. S6H35 2005
　572—dc22

　　　　　　　　　　　　　　　　　　　　　　　　　　　　　　2004028306

Printed in the United States of America

10　9　8

CONTENTS

PREFACE

This book is intended as a companion to *Thermodynamics and Kinetics for the Biological Sciences*, published in 2000. These two books are based on a course that has been given to first-year graduate students in the biological sciences at Duke University. These students typically do not have a strong background in mathematics and have not taken a course in physical chemistry. The intent of both volumes is to introduce the concepts of physical chemistry that are of particular interest to biologists with a minimum of mathematics. I believe that it is essential for all students in the biological sciences to feel comfortable with quantitative interpretations of the phenomena they are studying. Indeed, the necessity to be able to use quantitative concepts has become even more important with recent advances, for example, in the fields of proteomics and genomics. The two volumes can be used for a one-semester introduction to physical chemistry at both the first-year graduate level and at the sophomore-junior undergraduate level. As in the first volume, some problems are included, as they are necessary to achieve a full understanding of the subject matter.

I have taken some liberties with the definition of spectroscopy so that chapters on x-ray crystallography and mass spectrometry are included in this volume. This is because of the importance of these tools for understanding biological phenomena. The intent is to give students a fairly complete background in the physical chemical aspects of biology, although obviously the coverage cannot be as complete or as rigorous as a traditional two-semester course in physical chemistry. The approach is more conceptual than traditional physical chemistry, and many examples of applications to biology are presented.

I am indebted to my colleagues at Duke for their assistance in looking over parts of the text and supplying material. Special thanks are due to Professors

David Richardson, Lorena Beese, Leonard Spicer, Terrence Oas, and Michael Fitzgerald. I again thank my wife, Judy, who has encouraged, assisted, and tolerated this effort. I welcome comments and suggestions from readers.

GORDON G. HAMMES

CHAPTER 1

FUNDAMENTALS OF SPECTROSCOPY

INTRODUCTION

Spectroscopy is a powerful tool for studying biological systems. It often provides a convenient method for analysis of individual components in a biological system such as proteins, nucleic acids, and metabolites. It can also provide detailed information about the structure and mechanism of action of molecules. In order to obtain the maximum benefit from this tool and to use it properly, a basic understanding of spectroscopy is necessary. This includes a knowledge of the fundamentals of spectroscopic phenomena, as well as of the instrumentation currently available. A detailed understanding involves complex theory, but a grasp of the important concepts and their application can be obtained without resorting to advanced mathematics and theory. We will attempt to do this by emphasizing the physical ideas associated with spectral phenomena and utilizing a few of the concepts and results from molecular theory.

Very simply stated, spectroscopy is the study of the interaction of radiation with matter. Radiation is characterized by its energy, E, which is linked to the frequency, υ, or wavelength, λ, of the radiation by the familiar Planck relationship:

$$E = h\upsilon = hc/\lambda \tag{1-1}$$

Spectroscopy for the Biological Sciences, by Gordon G. Hammes
Copyright © 2005 John Wiley & Sons, Inc.

where c is the speed of light, 2.998×10^{10} cm/s (2.998×10^8 m/s), and h is Planck's constant, 6.625×10^{-27} erg-s (6.625×10^{-34} J-s). Note that $\lambda \upsilon = c$.

Radiation can be envisaged as an electromagnetic sine wave that contains both electric and magnetic components, as shown in Figure 1-1. As shown in the figure, the electric component of the wave is perpendicular to the magnetic component. Also shown is the relationship between the sine wave and the wavelength of the light. The useful wavelength of radiation for spectroscopy extends from x-rays, $\lambda \sim 1\text{–}100$ nm, to microwaves, $\lambda \sim 10^5\text{–}10^6$ nm. For biology, the most useful radiation for spectroscopy is in the ultraviolet and visible region of the spectrum. The entire useful spectrum is shown in Figure 1-2, along with the common names for the various regions of the spectrum. If

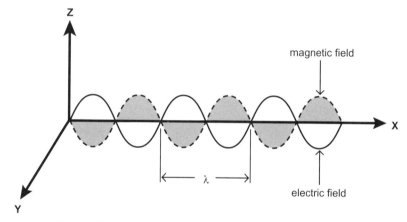

Figure 1-1. Schematic representation of an electromagnetic sine wave. The electric field is in the xz plane and the magnetic field in the xy plane. The electric and magnetic fields are perpendicular to each other at all times. The wavelength, λ, is the distance required for the wave to go through a complete cycle.

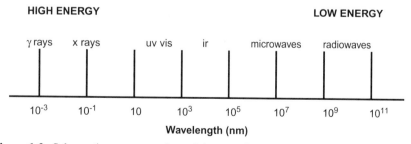

Figure 1-2. Schematic representation of the wavelengths associated with electromagnetic radiation. The wavelengths, in nanometers, span 14 orders of magnitude. The common names of the various regions also are indicated approximately (uv is ultraviolet; vis is visible; and ir is infrared).

radiation is envisaged as both an electric and magnetic wave, then its interactions with matter can be considered as electromagnetic phenomena, due to the fact that matter is made up of positive and negative charges. We will not be concerned with the details of this interaction, which falls into the domain of quantum mechanics. However, a few of the basic concepts of quantum mechanics are essential for understanding spectroscopy.

QUANTUM MECHANICS

Quantum mechanics was developed because of the failure of Newtonian mechanics to explain experimental results that emerged at the beginning of the 20th century. For example, for certain metals (e.g., Na), electrons are emitted when light is absorbed. This *photoelectric* effect has several nonclassical characteristics. First, for light of a given frequency, the kinetic energy of the electrons emitted is independent of the light intensity. The number of electrons produced is proportional to the light intensity, but all of the electrons have the same kinetic energy. Second, the kinetic energy of the photoelectron is zero until a threshold energy is reached, and then the kinetic energy becomes proportional to the frequency. This behavior is shown schematically in Figure 1-3, where the kinetic energy of the electrons is shown as a function

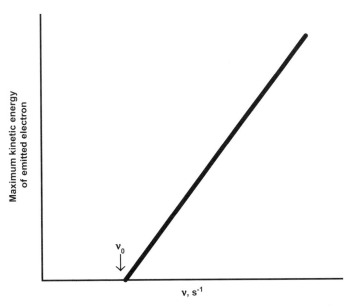

Figure 1-3. Schematic representation of the photoelectric effect. The maximum kinetic energy of an electron emitted from a metal surface when it is illuminated with light of frequency υ is shown. The frequency at which electrons are no longer emitted determines the work function, $h\upsilon_0$, and the slope of the line is Planck's constant (Eq. 1-2).

of the frequency of the radiation. An explanation of these phenomena was proposed by Einstein, who, following Planck, postulated that energy is absorbed only in discrete amounts of energy, $h\upsilon$. A photon of energy $h\upsilon$ has the possibility of ejecting an electron, but a minimum energy is necessary. Therefore,

$$\text{Kinetic Energy} = h\upsilon - h\upsilon_0 \qquad (1\text{-}2)$$

where $h\upsilon_0$ is the *work function* characteristic of the metal. This predicts that altering the light intensity would affect only the number of photoelectrons and not the kinetic energy. Furthermore, the slope of the experimental plot (Fig. 1-3) is h.

This explanation of the photoelectric effect postulates that light is corpuscular and consists of discrete photons characterized by a specific frequency. How can this be reconciled with the well-known wave description of light briefly discussed above? The answer is that both descriptions are correct— light can be envisioned either as discrete photons or a continuous wave. This wave-particle duality is a fundamental part of quantum mechanics. Both descriptions are correct, but one of them may more easily explain a given experimental situation.

About this point in history, de Broglie suggested this duality is applicable to matter also, so that matter can be described as particles or waves. For light, the energy is equal to the momentum, p, times the velocity of light, and by Einstein's postulate is also equal to $h\upsilon$.

$$E = h\upsilon = pc \qquad (1\text{-}3)$$

Furthermore, since $\lambda\upsilon = c$, $p = h/\lambda$. For macroscopic objects, $p = mv$, where v is the velocity and m is the mass. In this case, $\lambda = h/(mv)$, the de Broglie wavelength. These fundamental relationships have been verified for matter by several experiments such as the diffraction of electrons by crystals. The postulate of de Broglie can be extended to derive an important result of quantum mechanics developed by Heisenberg in 1927, namely the uncertainty principle:

$$\Delta p \, \Delta x \geq h/(2\pi) \qquad (1\text{-}4)$$

In this equation, Δp represents the uncertainty in the momentum and Δx the uncertainty in the position. The uncertainty principle means that it is not possible to determine the precise values of the momentum, p, and the position, x. The more precisely one of these variables is known, the less precisely the other variable is known. This has no practical consequences for macroscopic systems but is crucial for the consideration of systems at the atomic level. For example, if a ball weighing 100 grams moves at a velocity of 100 miles per hour (a good tennis serve), an uncertainty of 1 mile per hour in the speed gives $\Delta p \sim 4.4 \times$

10^{-2} kg m/sec and $\Delta x \sim 2 \times 10^{-33}$ m. We are unlikely to worry about this uncertainty! On the other hand, if an electron (mass = 9×10^{-28} g) has an uncertainty in its velocity of 1×10^8 cm/sec, the uncertainty in the position is about 1 Å, a large distance in terms of atomic dimensions. As we will see later, quantum mechanics has an alternative way of defining the position of an electron.

A second puzzling aspect of experimental physics in the late 1800s and early 1900s was found in the study of atomic spectra. Contrary to the predictions of classical mechanics, discrete lines at specific frequencies were observed when atomic gases at high temperatures emitted radiation. This can only be understood by the postulation of discrete energy levels for electrons. This was first explained by the famous Bohr atom, but this model was found to have shortcomings, and the final resolution of the problem occurred only when quantum mechanics was developed by Schrödinger and Heisenberg in the late 1920s. We will only consider the development by Schrödinger, which is somewhat less complex than that of Heisenberg.

Schrödinger postulated that all matter can be described as a wave and developed a differential equation that can be solved to determine the properties of a system. Basically, this differential equation contains two important variables, the kinetic energy and the potential energy. Both of these are well-known concepts from classical mechanics, but they are redefined in the development of quantum mechanics. If the wave equation is solved for specific systems, it fully explains the previously puzzling results. Energy is quantized, so discrete energy levels are obtained. Furthermore, a consequence of quantum mechanics is that the position of a particle can never be completely specified. Instead, the probability of finding a particle in a specific location can be determined, and the average position of a particle can be calculated. This probabilistic view of matter is in contrast to the deterministic character of Newtonian mechanics and has sparked considerable philosophic debate. In fact, Einstein apparently never fully accepted this probabilistic view of nature. In addition to the above concepts, quantum mechanics also permits quantitative calculations of the interaction of radiation with matter. The result is the specification of rules that ultimately determine what is observed experimentally. We will make use of these rules without considering the details of their origin, but it is important to remember that they stem from detailed quantum mechanical calculations.

PARTICLE IN A BOX

As an example of a simple quantum mechanical result that leads to quantization of energy levels, we consider a particle of mass m moving back and forth in a one-dimensional box of length L. This actually has some practical application. It is a good model for the movement of pi electrons that are delocalized over a large part of a molecule, for example, biological molecules such as carotenoids, hemes, and chlorophyll. This is not an ordinary box because inside

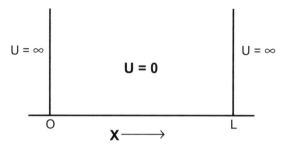

Figure 1-4. Quantum mechanical model for a particle in a one-dimensional box of length L. The particle is confined to the box by setting the potential energy equal to 0 inside the box and to ∞ outside of the box.

the box, the potential energy of the system is 0, whereas outside of the box, the potential energy is infinite. This is depicted in Figure 1-4. The Schrödinger equation in one dimension is

$$-\frac{h^2}{8m\pi^2}\frac{d^2\psi_n}{dx^2} + U = E_n\psi_n \tag{1-5}$$

where Ψ_n is the wave function, x is the position coordinate, U is the potential energy, and E_n is the energy associated with the wave function Ψ_n. Since the potential walls are infinitely high, the solution to this equation outside of the box is easy—there is no chance the particle is outside the box so the wave function must be 0. Inside the box, U = 0, and Eq. 1-5 can be easily solved. The solution is

$$\psi_n = A \sin bx \tag{1-6}$$

where A and b are constants. At the ends of the box, Ψ must be zero. This happen when sin $n\pi$ = 0 and n is an integer, so b must be equal to $n\pi/L$. This causes Ψ_n to be 0 when x = 0 and x = L for all integral values of n. To evaluate A, we introduce another concept from quantum mechanics, namely that the probability of finding the particle in the interval between x and x + dx is $\Psi^2 dx$. Since the particle must be in the box, the probability of finding the particle in the box is 1, or

$$\int_0^L \psi_n^2 dx = \int_0^L A^2 \sin^2(n\pi x/L)dx = 1 \tag{1-7}$$

Evaluation of this integral gives $A = \sqrt{2/L}$ Thus the final result for the wave function is

$$\psi_n = \sqrt{2/L} \sin(n\pi x/L) \tag{1-8}$$

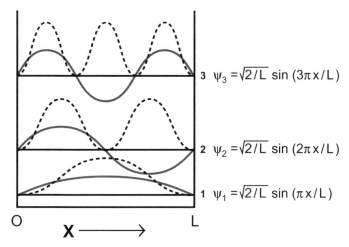

3 $\psi_3 = \sqrt{2/L} \, \sin(3\pi x/L)$

2 $\psi_2 = \sqrt{2/L} \, \sin(2\pi x/L)$

1 $\psi_1 = \sqrt{2/L} \, \sin(\pi x/L)$

O   **X** $\longrightarrow$ L

Figure 1-5. Wave functions, Ψ, for the first three energy levels of the particle in a box (Eq. 1-8). The dashed lines show the probability of the finding the particle at a given position x, Ψ^2.

Obviously n cannot be 0, as this would predict that there is no probability of finding the particle in the box, but n can be any integer. The wave functions for a few values of n are shown in Figure 1-5. Basically Ψ_n is a sine wave, with the "wavelength" decreasing as n increases. (More advanced treatments of quantum mechanics use the notation associated with complex numbers in discussing the wave equation and wave functions, but this is beyond the scope of this text.)

To determine the energy of the particle, all we have to do is put Eq. 1-8 back into Eq. 1-5 and solve for E_n. The result is

$$E_n = (h^2 n^2)/(8mL^2) \qquad (1-9)$$

Thus, we see that the energy is quantized, and the energy is characterized by a series of energy levels, as depicted in Figure 1-6. Each energy level, E_n, is associated with a specific wave function, Ψ_n. Notice that the energy levels would be very widely spaced for a very light particle such as an electron, but would be very closely spaced for a macroscopic particle. Similarly, the smaller the box, the more widely spaced the energy levels. For a tennis ball being hit on a tennis court, the ball is sufficiently heavy and the court (box) sufficiently big so that the energy levels would be a continuum for all practical purposes. The uncertainty in the momentum and position of the ball cannot be blamed on quantum mechanics in this case! The particle in a box illustrates how quantum mechanics can be used to calculate the properties of systems and how quantization of energy levels arises. The same calculation can be easily done for a three-dimensional box. In this case, the energy states are the sum

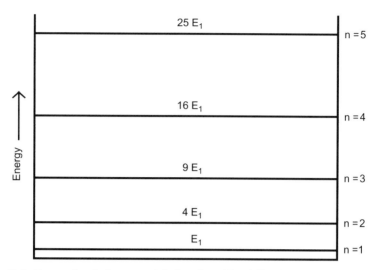

Figure 1-6. Energy levels for a particle in a box (Eq. 1-9). The energy levels are n^2E_1 where E_1 is the energy when $n = 1$.

of three terms identical to Eq. 1-9, but with each of the three terms having a different quantum number.

The quantum mechanical description of matter does not permit determination of the precise position of the particle to be determined, a manifestation of the Heisenberg uncertainty principle. However, the probability of finding the particle within a given segment of the box can be calculated. For example, the probability of finding the particle in the middle of the box, that is, between L/4 and 3L/4 for the lowest energy state is

$$\int_{L/4}^{3L/4} \psi_1^2 dx = (2/L) \int_{L/4}^{3L/4} \sin^2(\pi x/L)dx$$

Evaluation of this integral gives a probability of 0.82. The probability of finding the particle within the middle part of the box is independent of L, the size of the box, but does depend on the value of the quantum number, n. For the second energy level, n = 2, the probability is 0.50. The probability of finding the particle at position x in the box is shown as a dashed line for the first three energy levels in Figure 1-5.

An important result of quantum mechanics is that not only do molecules exist in different discrete energy levels, but the interaction of radiation with molecules causes shifts between these energy levels. If energy or radiation is *absorbed* by a molecule, the molecule can be raised to a higher energy state, whereas if a molecule loses energy, radiation can be *emitted*. For both cases, the change in energy is related to the radiation that is absorbed or emitted by a slight modification of Eq. 1-1, namely the change in energy state of the molecule, ΔE, is

$$\Delta E = h\upsilon = hc/\lambda \tag{1-10}$$

The change in energy, ΔE, is the difference in energy between specific energy levels of the molecule, for example, $E_2 - E_1$ where 1 and 2 designate different energy levels. It is important to note that since the energy is quantized, the light emitted or absorbed is always a specific single frequency. Equation 1-10 can be applied to the particle in a box for the particle dropping from the $n + 1$ energy level to the n energy level:

$$\Delta E = \frac{h^2}{8mL^2}\left[(n+1)^2 - n^2\right] = \frac{h^2}{8mL^2}(2n+1) = hc/\lambda \tag{1-11}$$

If the particle is assumed to be an electron moving in a molecule $20\,\text{Å}$ long and $n = 10$, then $\lambda \sim 600\,\text{nm}$. This wavelength is in the visible region and has been observed for π electrons that are highly delocalized in molecules.

In practice, energy levels are sometimes so closely spaced that the frequencies of light emitted appear to create a continuum of frequencies. This is a shortcoming of the experimental method—in reality the frequencies emitted are discrete entities. The particle in a box is a rather simple application of quantum mechanics, but it illustrates several important points that also are found in more complex calculations for molecular systems. First, the system can be described by a wave function. Second, this wave function permits determination of the probability of important characteristics of the system, such as positions. Finally, the energy of the system can be calculated and is found to be quantized. Moreover, the energy can only be absorbed or emitted in quantized packages characterized by specific frequencies. Quantum mechanical calculations also tell us what conditions are necessary for energy to be emitted or absorbed by a molecule. These calculations tell us *whether* radiation will be emitted or absorbed and what quantized packets of energy are available. We will only utilize the results of these calculations and will not be concerned with the details of the interactions between light and molecules other than the above concepts.

PROPERTIES OF WAVES

It is useful to consider several additional aspects of light waves in order to understand better some of the experimental methods that will be discussed later. Thus far we have considered light to be a periodic electromagnetic wave in space that could be characterized, for example, by a sine function:

$$I = I_0 \sin(2\pi x/\lambda) \tag{1-12}$$

Here I is the magnetic or electric field, I_0 is the maximum value of the electric or magnetic field, x is the distance along the x axis and λ is the wavelength.

A light wave can also be periodic in time, as illustrated in Figure 1-7. In this case:

$$I = I_0 \sin 2\pi \upsilon t = I_0 \sin \omega t \qquad (1\text{-}13)$$

Now, I is the light intensity, I_0 is the maximum light intensity, υ is the frequency in s^{-1}, as defined in Figure 1-7, ω is the frequency in radians ($\omega = 2\pi\upsilon$), and t is the time. The velocity of the propagating wave is $\lambda\upsilon$, which in the case of electromagnetic radiation is the speed of light, that is, $\lambda\upsilon = c$. If light of the same frequency and maximum amplitude from two sources is combined, the two sine functions will be added. If the two light waves start with zero intensity at the same time (t = 0), the two waves add and the intensity is doubled. This is called *constructive interference*. If the two waves are combined with one of the waves starting at zero intensity and proceeding to positive values of the sine function, whereas the other begins at zero intensity and proceeds to negative values, the two intensities cancel each other out. This is called *destructive interference*. Obviously it is possible to have cases in between these two extremes. In such cases, a phase difference is said to exist between the two waves. Mathematically this can be represented as

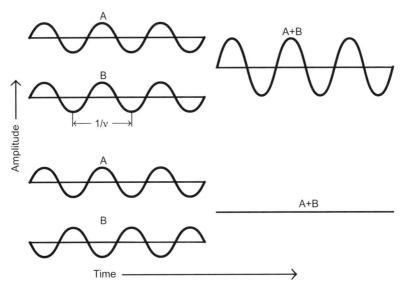

Figure 1-7. Examples of constructive and destructive interference. Constructive interference: when the upper two wave forms of equal amplitude and a phase angle of 0° (or integral multiples of 2π) are added (*left*), a sine wave with twice the amplitude and the same frequency results (*right*). Destructive interference: when the lower two wave forms are added (*left*), the amplitudes of the two waves cancel (*right*). The phase angle in this case is 90° (or odd integral multiples of $\pi/2$).

$$I = I_0 \sin(\omega t + \delta) \tag{1-14}$$

where δ is called a phase angle and can be either positive or negative. When many different waves of the same frequency are combined, the intensity can always be described by such a relationship. These phenomena are shown schematically in Figure 1-7.

A standard way of carrying out spectroscopy is to apply continuous radiation, and then look at the intensity of the radiation after it has passed through the sample of interest. The intensity is then determined as a function of the frequency of the radiation, and the result is the absorption spectrum of the sample. The color of a material is determined by the wavelength of the light absorbed. For example, if white light shines on blood, blue/green light is absorbed so that the transmitted light is red. Several examples of absorption spectra are shown in Figure 1-8. We will consider why and how much the sample absorbs light a bit later, but you are undoubtedly already familiar with the concept of an absorption spectrum.

The use of continuous radiation is a useful way to carry out an experiment, but there is an interesting mathematical relationship that permits a different approach to the problem. This mathematical operation is the *Fourier transform*. The principle of a Fourier transform is that if the frequency dependence of the intensity, $I(\upsilon)$, can be determined, it can be transformed into a new function, $F(t)$, that is a function of the time, t. Conversely, $F(t)$ can also be converted to $I(\upsilon)$. Both of these functions contain the same information. Why then are these transformations advantageous? It can be quite time consuming to determine $I(\upsilon)$, but a short pulse of radiation can be applied very quickly. Basically what this transformation means is that looking at the response of the system to application of a pulse of radiation, such as shown as in Figure 1-9, is equivalent to looking at the response of the system to sine wave radiation at many different frequencies. In other words, a square wave is mathematically equivalent to adding up many sine waves of different frequency, and vice versa. This is shown schematically in Figure 1-9 where the addition of sine waves with four different frequencies produces a periodic "square" wave. The larger the number of sine waves added, the more "square" the wave becomes. In mathematical terms, a square wave can be represented as an infinite series of sine functions, a Fourier series.

The mathematical equivalence of timed pulses and continuous waves of many different frequencies has profound consequences in determining the spectroscopic properties of materials. In many cases, the use of pulses permits thousands of experiments to be done in a very short time. The results of these experiments can then be averaged, producing a far superior frequency spectrum in a much shorter time than could be determined by continuous wave methods. In later chapters, we will be dealing both with continuous wave spectroscopy and Fourier transform spectroscopy. It is important to remember that

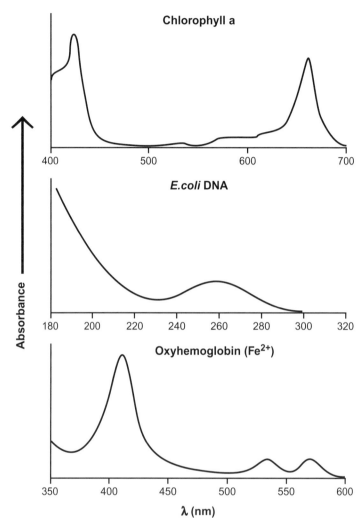

Figure 1-8. Absorption of light by biological molecules. The absorbance scale is arbitrary and the wavelength, λ, is in nanometers. Chlorophyll a solutions absorb blue and red light and are green in color. DNA solutions absorb light in the ultraviolet and are colorless. Oxyhemoglobin solutions absorb blue light and are red in color.

both methods give identical results. The method of choice is that one that produces the best data in the shortest time, and in some cases at the lowest cost.

With this brief introduction to the underlying theoretical principles of spectroscopy, we are ready to proceed with consideration of specific types of spectroscopy and their application to biological systems.

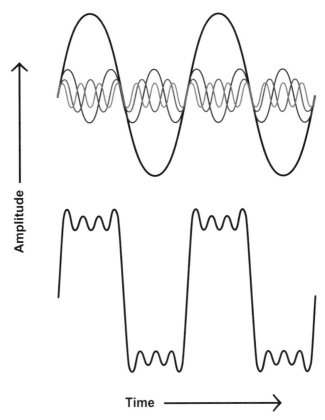

Figure 1-9. The upper part of the figure shows sine waves of four different frequencies, and the lower part of the figure is the sum of the sine waves, which approximates a square wave pulse of radiation. When sine waves of many more frequencies are included, the time dependence becomes a pulsed square wave. This figure illustrates that the superposition of multiple sine waves is equivalent to a square wave pulse and vice versa. This equivalency is the essence of Fourier transform methods. Copyright by Professor T. G. Oas, Duke University. Reproduced with permission.

REFERENCES

The topics in this chapter are discussed in considerably more depth in a number of physical chemistry textbooks, such as those cited below.

1. I. Tinoco Jr., K. Sauer, J. C. Wang, and J. D. Puglisi, *Physical Chemistry: Principles and Applications in Biological Sciences*, 4th edition, Prentice Hall, Englewood, NJ, 2002.

2. R. J. Silbey, R. A. Alberty, and M. G. Bawendi, *Physical Chemistry*, 4th edition, John Wiley & Sons, New York, 2004.

3. P. W. Atkins and J. de Paula, *Physical Chemistry*, 7th edition, W. H. Freeman, New York, 2001.

4. R. S. Berry, S. A. Rice, and J. Ross, *Physical Chemistry*, 2nd edition, Oxford University Press, New York, 2000.

5. D. A. McQuarrie and J. D. Simon, *Physical Chemistry: A Molecular Approach*, University Science Books, Sausalito, CA, 1997.

PROBLEMS

1.1. The energies required to break the C–C bond in ethane, the "triple bond" in CO, and a hydrogen bond are about 88, 257, and 4 kcal/mol. What wavelengths of radiation are required to break these bonds?

1.2. Calculate the energy and momentum of a photon with the following wavelengths: 150 pm (x ray), 250 nm (ultraviolet), 500 nm (visible), and 1 cm (microwave).

1.3. The maximum kinetic energy of electrons emitted from Na at different wavelengths was measured with the following results.

λ (Å)	Max Kinetic Energy (electron volts)
4500	0.40
4000	0.76
3500	1.20
3000	1.79

Calculate Planck's constant and the value of the work function from these data. (1 electron volt = 1.602×10^{-19} J)

1.4. Calculate the de Broglie wavelength for the following cases:

a. An electron in an electron microscope accelerated with a potential of 100 kvolts.

b. A He atom moving at a speed of 1000 m/s.

c. A bullet weighing 1 gram moving at a speed of 100 m/s.

Assume the uncertainty in the speed is 10%, and calculate the uncertainty in the position for each of the three cases.

1.5. The particle in a box is a useful model for electrons that can move relatively freely within a bonding system such as π electrons. Assume an electron is moving in a "box" that is 50 Å long, that is, a potential well with infinitely high walls at the boundaries.

a. Calculate the energy levels for n = 1, 2, and 3.

b. What is the wavelength of light emitted when the electron moves from the energy level with n = 2 to the energy level with n = 1?

    c. What is the probability of finding the electron between 12.5 and 37.5 Å for $n = 1$.

1.6. Sketch the graph of I versus t for sine wave radiation that obeys the relationship $I = I_0 \sin(\omega t + \delta)$ for $\delta = 0$, $\pi/4$, $\pi/2$, and π.

Plot the sum of the sine waves when the sine wave for $\delta = 0$ is added to that for $\delta = 0$ or $\pi/4$, or $\pi/2$, or π. This exercise should provide you with a good understanding of constructive and destructive interference.

Do your results depend on the value of ω? Briefly discuss what happens when waves of different frequency are added together.

CHAPTER 2

X-RAY CRYSTALLOGRAPHY

INTRODUCTION

The primary tool for determining the atomic structure of macromolecules is the scattering of x-ray radiation by crystals. Strictly speaking this is not considered spectroscopy, even though it involves the interaction of radiation with matter. Nevertheless, the importance of this tool in modern biology makes it mandatory to have some understanding of how macromolecular structures are determined with x-ray radiation. The basic principles of the methodology will be discussed without delving into the details of how structures are determined. We then consider some of the important results that have been obtained, as well as how they can be used to understand the function of macromolecules in biological systems.

Although considerable progress has been made in determining the structures of macromolecules that are not strictly crystalline, we will concentrate on the structure of macromolecules from high-quality crystals. High-quality crystals have molecules arranged in a regular array, and this regular array, or lattice, serves as a scattering surface for x rays.

Why are x rays used to determine molecular structures? Ordinary objects can be easily seen with visible light, and very small objects can be seen in microscopes with a high-quality lens that converts the electromagnetic waves associated with light into an image. However, the resolution of a light microscope is limited by the wavelength of light that is used. Distances cannot be

Spectroscopy for the Biological Sciences, by Gordon G. Hammes
Copyright © 2005 John Wiley & Sons, Inc.

resolved that are significantly shorter than the wavelength of the light that illuminates the object. The wavelength of visible light is thousands of angstroms, whereas distances within molecules are approximately angstroms.

The obvious answer to this resolution problem is to use radiation that has a much shorter wavelength than visible light, namely x rays which have wavelengths in the angstrom region. In principle, all that is needed is a lens that will convert the scattering from a molecule into an image. What does this entail? We have previously shown that light has an amplitude and a periodicity with respect to time and distance, or a phase. A lens takes both the amplitude and phase information and converts it into an image. Unfortunately, a lens does not exist that will carry out this function for x-ray radiation. Instead what can be measured is the amplitude of the scattered radiation. Methods are then needed to obtain the phase and ultimately the molecular structure. The principles behind this methodology are given below, without presenting the underlying mathematical complications. More advanced texts should be consulted for the mathematical details (1, 2).

SCATTERING OF X RAYS BY A CRYSTAL

X rays are produced by bombarding a target (typically a metal) with high-energy electrons. If the energy of the incoming electron is sufficiently large, it will eject an electron from an inner orbital of the target. A photon is emitted when an outer orbit electron moves into the vacated inner orbital. For typical targets, the wavelengths of the photons are tenths of angstrom to several angstroms. In a normal laboratory experiment, x rays are produced by special electronic tubes. However, more intense short-wavelength sources can be obtained using high-energy electron accelerators (synchrotrons), and these sources are often used to obtain high-resolution structures. The intensity of x rays from a synchrotron is thousands of times greater than a conventional source and radiation of different wavelengths can be selected. Smaller crystals can be used for structure determinations with synchrotron radiation than with conventional sources, which is a considerable advantage.

Crystals can be considered as a regular array of molecules, or scattering elements. Only seven symmetry arrangements of crystals are possible. For example, one possibility is a simple cube. In this case, the three axes are equal in length and are 90° with respect to each other. Other lattices can be generated by having angles other than 90° and unequal sides of the geometric figure. A summary of the seven crystal systems is given in Table 2.1.

The crystal type does not give a complete picture of the possible arrangements of atoms within the crystal, that is, the lattice. The lattice is an infinite array of points (atoms) in space. A. Bravais showed that all lattices fall within 14 types, the *Bravais lattices*. For example, for a cubic crystal, three lattices are possible: one with an atom at each corner of the cube, one with an additional atom at the center of the cube (body-centered cube), and one with an addi-

TABLE 2-1. The Seven Crystal Systems

System	Axes	Angles
Cubic	$a = b = c$	$\alpha = \beta = \gamma = 90°$
Rhombohedral	$a = b = c$	$\alpha = \beta = \gamma$
Tetragonal	$a = b; c$	$\alpha = \beta = \gamma = 90°$
Hexagonal	$a = b; c$	$\alpha = \beta = 90°; \gamma = 120°$
Orthorhombic	$a; b; c$	$\alpha = \beta = \gamma = 90°$
Monoclinic	$a; b; c$	$\alpha = \gamma = 90°; \beta$
Triclinic	$a; b; c$	$\alpha; \beta; \gamma$

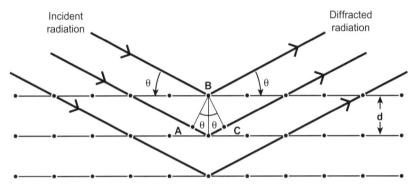

Figure 2-1. Diffraction of x-ray radiation by a crystal lattice. The parallel lines are planes of atoms separated by a distance, d, and the radiation impinges on the crystal at an angle θ. This diagram can be used to derive the Bragg condition for maximum constructive interference as described in the text (Eq. 2-1).

tional atom in the center of each face of the cube (face-centered cube). Determination of crystal type and lattice classification is important for the subsequent analysis of the x-ray scattering data.

W. L. Bragg showed that the scattering of x rays from a crystal can be described as the scattering from parallel planes of molecules, as illustrated in Figure 2-1. If the incident beam of x rays is at an angle θ with respect to the molecular plane, then it will be scattered at an angle θ. This is called *elastic* scattering and assumes that the radiation does not lose or absorb energy in the scattering process. In reality, some *inelastic* scattering occurs in which energy can be gained or lost, but this effect can be neglected. From the diagram in Figure 2-1, it is clear that scattering will occur for each plane and scattering center with the same incident and exit angle. From a wave standpoint, the radiation will be scattered from each plane, and the radiation from each plane will have a different phase as it exits since each wave will have traveled a different distance depending on the depth of the plane in Figure 2-1. However, if the wavelength of radiation is such that the difference in path length traveled by the beams from different planes is equal to the wavelength or an integer multiple of the wavelength, then the two waves will be in phase and constructive interference will occur, that is, the intensity of the radiation will be at a maximum. The phenomena of constructive and destructive interference have been discussed in Chapter 1. Figure 1-7 illustrates these phenomena when the time dependence of an electromagnetic wave is considered. The same analysis is applicable for the propagation of a wave in space (Fig. 1-1 and Eq. 1-12), as illustrated in Figure 2-2.

The condition for maximum constructive interference can be calculated by reference to Figure 2-1:

$$AB + BC = n\lambda$$

where n is an integer and λ is the wavelength. The distances AB and BC are equal to $d\sin\theta$ where d is the distance between planes. Thus, the Bragg equation is

$$n\lambda = 2d\sin\theta \qquad (2\text{-}1)$$

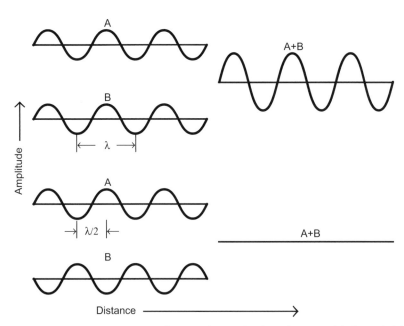

Figure 2-2. Illustration of constructive and destructive interference. This figure is identical to Figure 1-7, except that now the abscissa is distance rather than time. The upper two waves are in phase, which means they differ by integral multiples of the wavelength, λ, and constructive interference occurs. The bottom two waves differ by $\lambda/2$ in phase, and destructive interference occurs.

The Bragg equation gives the condition for diffraction so that if a crystal is rotated in a beam of x rays, the scattering pattern is a series of intensity maxima. A real crystal is more complex than a set of parallel planes of point-scattering sources. In fact, multiple planes exist, and a molecule is not a simple point scatterer. Electrons in atoms are the scatterers, and each atom has a different effectiveness as a scatterer. Consequently, when an experiment is carried out, a set of diffraction maxima are observed of different intensities. A schematic representation of the experimental setup is shown in Figure 2-3. Either the crystal or detector, or both, are rotated to obtain the scattering intensity at various angles. The diffraction pattern has a strong peak at the center, the unscattered beam, and a radial distribution around the center, corresponding to different planes and values of n in Eq. 2-1.

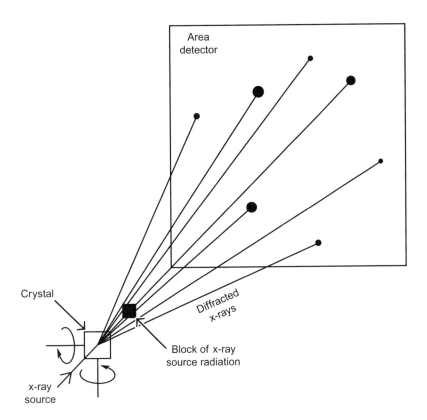

Figure 2-3. Experimental setup for measuring x-ray diffraction from crystals. The crystal can be rotated around both perpendicular axes, and the diffraction pattern is measured on an area detector or film. The incident x-ray radiation is prevented from hitting the area detector by insertion of a beam blocker between the crystal and the detector.

The Bragg equation can be used to derive the minimum spacing of planes that can be resolved for x rays of a given wavelength. Since the maximum value of $\sin\theta$ is 1, $d_{min} = \lambda/2$.

STRUCTURE DETERMINATION

The analysis of a diffraction pattern is considerably more difficult than suggested above. The scattering intensity depends on the scattering effectiveness of the individual atoms and the phase of the wave from each scattering source. The structure factor, F, for each plane can be defined as the sum of the structure factors for individual atoms, f_i, times a phase factor, α_i, for each atom:

$$F = \sum_i f_j a_i \tag{2-2}$$

The intensity of scattered radiation is proportional to the absolute value of the amplitude of the structure factor. The structure factor can be calculated if the atomic coordinates are known. Note that Eq. 2-2 is analogous to the general description of electromagnetic waves developed in Chapter 1, namely an amplitude is multiplied by an angular dependence (sin and/or cos).

In proceeding to determination of the structure, it should be remembered that x rays are scattered by electrons in atoms so that representing a crystal as a series of point atoms is not a good picture of the real situation. Instead, a more realistic formulation is to represent the structure factor as an integral over a continuous distribution of electron density. The electron density is a function of the coordinates of the scattering centers, the atoms, and has a maximum around the position of each atom. What is desired in practice is to convert the measured structure factors into atomic coordinates. This is done by taking the Fourier transform of Eq. 2-2. In this case, the Fourier transform takes the structure factors, which are functions of the electron density, and inverts the functional dependence so that the electron density is expressed as a function of the structure factors. (This is analogous to the discussion in Chapter 1 in which the frequency and time dependence are interchanged by Fourier transforms.) This seems straightforward from the standpoint of the mathematics, but the problem is that the actual structure factors contain both amplitudes and phases. Only the amplitude, or to be more precise, its square, can be directly derived from the measured intensity of the diffracted beam. The phase factor must be determined before a structure can be calculated.

Two methods are commonly used to solve the phase problem for macromolecules. One of these is Multiple-wavelength Anomalous Dispersion or MAD. Thus far, we have discussed x-ray scattering in terms of *coherent* scattering where the frequency of the radiation is different than the frequency of oscillation of electrons in the atoms. However, x-ray frequencies are available that match the frequency of oscillation of the electrons in some atoms, giving rise to *anomalous* scattering that can be readily discerned from the wavelength dependence of the scattering. The feasibility of this approach is made possible by the use of synchrotron radiation since crystal monochromaters can be used to obtain high-intensity x rays at a variety of wavelengths. Anomalous scattering is usually observed with relatively heavy atoms (e.g., metals or iodine) that can be inserted into the macromolecular structure. The observation of anomalous scattering for specific atoms allows these atoms to be located in the structure, thus providing the phase information that is required to obtain the complete structure.

Another frequently used method for determining phases in macromolecules is *isomorphous replacement*. With this method a few heavy atoms are put into the structure, for example, metal ions. Since scattering is proportional to the square of the atomic number, the enhanced scattering due to the heavy atoms can be easily seen, and the positions of the heavy atoms determined. This is done by looking at the difference in structure factors between the native structure and its heavy atom derivative. Of course, it is essential that the isomorphous replacement not significantly alter the structure of the molecule being studied. We will not delve into the details of the methodology here.

Thus far the assumption has been made that the x-ray radiation is monochromatic. An alternative is to use a broad spectrum of radiation ("white" radiation). The premise is that whatever the orientation of the crystal, one of the wavelengths would satisfy the Bragg condition. In fact, this was the basis for the first x-ray diffraction experiments carried out by Max von Laue in 1912, and this approach is called Laue diffraction. As might be expected, Laue diffraction patterns can be very complex and difficult to interpret as diffraction patterns from multiple wavelengths are observed simultaneously. The advantage is that a large amount of structural information can be obtained in a very short time. A Laue diffraction pattern of DNA polymerase from *Bacillus stearothermophilus* obtained with an area detector and synchrotron radiation is shown in Figure 2-4. Although most structural determinations use monochromatic radiation, the use of Laue diffraction has increased significantly in recent years.

What limits the precision of a structure determination? Real crystals are not infinite arrays of planes. They have imperfections so that the diffraction pattern is strong near the center and becomes weaker radially from the center, as the reflections from planes closer together become important. The precision of the distances in the final structure is limited by how many reflections are observed. The better the order in the crystal, the further from the center of the diffraction pattern that reflections can be seen. This problem can only be cured by obtaining better crystals. A minor problem is that atoms have thermal motion, so that an inherent uncertainty in the position exists. This can be helped by working at low temperatures, which sometimes will lock otherwise mobile structures into one conformation. Finally, it should be noted that in some cases portions of the macromolecule may not be well defined because of intrinsic disorder, that is, more than one arrangement of the atoms occurs in the crystals.

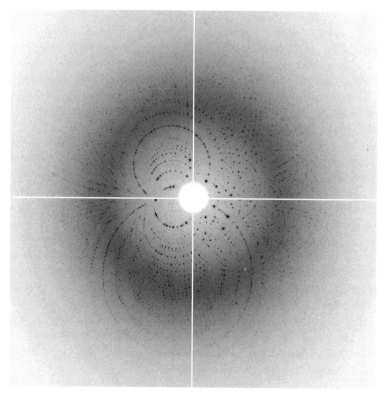

Figure 2-4. Laue diffraction pattern obtained with synchrotron radiation and an area detector. The protein crystal is DNA polymerase from *Bacillus stearothermophilus*. Copyright by Professor L. S. Beese, Duke University. Reproduced with permission.

The spatial resolution of a structure is usually given in terms of the distances that can be distinguished in the diffraction pattern. For good structures, this is typically 2 Å or better. However, this is not the uncertainty in the positions of atoms within the structure. In solving the structure of a macromolecule, other information is used from known structures of small molecules, for example, bond distances (C–C, etc.) and bond angles. For high-quality structures, the uncertainty in atomic positions is tenths of angstroms.

At the present time, the primary difficulty in determining the molecular structures of proteins and nucleic resides is obtaining good crystals, that is, crystals that give good diffraction patterns. If good diffraction patterns can be obtained, including isomorphous replacements or anomalous scattering atoms, computer programs are available to help derive the structure. However, as structures become larger and larger, obtaining good crystals and the data analysis itself are both a challenge.

Crystal structures represent a static picture of the equilibrium structure so that the correlation with biological function must be approached with caution.

Furthermore, the packing of molecules within a crystal can alter the structure relative to that in solution. This is usually true only for atoms on the surface of the molecule so that the core structures, such as active sites, are normally an accurate reflection of the biologically active species. Recent applications of synchrotron radiation have permitted the time evolution of structures to be studied using Laue diffraction (3). Although such experiments are exceedingly difficult to carry out and interpret, in principle it is possible to observe the structure of biological molecules as they function.

NEUTRON DIFFRACTION

High-velocity thermal neutrons are generated by atomic reactors. If they are slowed down by collisions, typically with D_2O, then their energy corresponds to a wavelength of about 1 Å. The scattering of neutrons by atoms is quite different than the scattering of x rays. In the latter case, the scattering is by electrons whereas in the former case the scattering is by the nuclei. Moreover, the scattering of neutrons is not easily related to atomic number.

A major difference between neutron and x-ray scattering is that hydrogen is a very effective scattering center for neutrons but is relatively ineffective for x rays. Consequently, the positions of hydrogen atoms are difficult to obtain from x-ray scattering but can be readily found through neutron scattering. D_2O scatters quite differently than H_2O: their scattering factors have opposite signs, that is, they are out of phase with respect to each other. The scattering factors for proteins and nucleic acids usually fall somewhere in between so that mixtures of D_2O and H_2O can be used as "contrast" agents. Appropriate mixtures can be used to effectively make certain macromolecules "invisible" to neutron scattering. For example, if a protein–nucleic acid interaction is being studied, the appropriate solvent mixture can make either the protein or the nucleic acid "invisible." Historically, this property was important for mapping the structure of the ribosome.

Neutron scattering studies are relatively rare because high-flux neutron sources are small in number. Nevertheless, neutron diffraction can be a useful tool in elucidating macromolecular structure.

NUCLEIC ACID STRUCTURE

Probably the most exciting structure determination of a biological molecule was that for B-DNA deduced by James Watson and Francis Crick. This structure was, in fact, based on diffraction patterns from fibers, rather than crystals. The familiar right-handed double helix is shown in Figure 2-5 (see color plates). In this structure two polynucleotide chains with opposite orientations coil around an axis to form the double helix. The purine and pyrimidine bases from different chains hydrogen bond in the interior of the double helix, with adenine (A) hydrogen bonding to thymine (T) and guanine (G) hydrogen bonding to cytosine (C). The former pair has two hydrogen bonds and the latter three hydrogen bonds. Hence, the G-C pair is significantly more stable than the A-T pair. Phosphate and deoxyribose are on the outside of the double helix and interact favorably with water. The double helical structure contains two obvious grooves, the major and minor grooves. The major groove is wider and deeper than the minor groove. These grooves arise because the glycosidic bonds of a base pair are not exactly opposite each other.

We will not dwell on the biological function of DNA except to note that DNA must separate during the replication process, and the stability of various DNA depends on the base composition of the DNA. At sufficiently high temperatures, all DNA structures are destroyed and the two polynucleotide chains separate.

Extensive x-ray studies have shown that other forms of helical DNA exist. If the relative humidity is reduced below about 75%, A-DNA forms. It also is a right-handed double helix of antiparallel strands, but a puckering of the sugar rings causes the bases to be tilted away from the normal of the axis. A-DNA is shorter and wider than B-DNA. This structure is found in biology for some double-stranded regions of RNA and in RNA-DNA hybrids.

A third type of DNA has been found that is a left-handed helix, Z-DNA. Z-DNA is elongated relative to B-DNA and has more unfavorable electrostatic interactions. This structure has been observed with specific short oligonucleotides at high salt concentrations. The biological significance of this structure is not clear, but its occurrence demonstrates that quite different structures can exist for DNA.

RNA has more diverse structures than DNA, in keeping with its diverse biological functions that include its role as messenger RNA (mRNA) during transcription, as transfer RNA (tRNA) in protein synthesis, and as ribozymes in catalysis. In addition, RNA is found in ribosomes and other ribonucleic proteins. Typically RNA does not form a double helix from two separate chains, as DNA does. However, the same base pairing rules as found for DNA cause internal helices to be formed within an RNA molecule. In fact, the structures of many RNA molecules are inferred in this way. In RNA, uracil (U) is found rather than thymine, as in DNA.

The first structure of an RNA deduced by x-ray crystallography was that of yeast phenylalanine tRNA (4). This structure is shown in Figure 2-6 (see color plates). The L-shaped molecule is typical of tRNAs. The hydrogen bonding network is shown in this structure. The acceptor stem (upper right-hand corner) is where the amino acid is linked to form the aminoacyl-tRNA. The amino acid is transferred to the growing protein chain during protein synthesis. The anticodon, which specifies the amino acid to be added to a protein during synthesis, is at the end of the long arm of the L. It pairs with a specific mRNA that is the genetic information for the amino acid. The structure of many other tRNA are now known and are quite similar.

For many years, a central dogma of biochemistry was that all physiological reactions are catalyzed by enzymes that are proteins. However, now many reactions are known that are catalyzed by RNA. These catalytic RNA (ribozymes) are particularly important in splicing and maturation of RNA. In some cases, the ribozyme cleaves other RNA whereas in other cases it undergoes self-cleavage. Ribosomal RNA also plays a catalytic role in the formation of peptide bonds. Although there is little doubt that RNA functions as a catalyst physiologically, they are not as efficient as enzymes. In some cases, such as self-cleavage, the reaction is not truly catalytic as multiple turnovers cannot occur. Many ribozymes require a protein to function efficiently, and even in cases where true catalysis occurs, it is slow relative to typical enzymatic reactions.

The structures of several ribozymes have been determined (5). The first structure that was determined was that of "hammerhead" ribozyme (6). This ribozyme was first discovered as a self-cleaving RNA associated with plant viruses. The minimal structure necessary for catalysis contains three short helices and a universally conserved junction. The structure is shown in Figure 2-7, along with an indication of the cleavage site in the RNA chain. The three helices form a Y-shaped structure, with helix II and III essentially in line, and helix I at a sharp angle to helix II. Distortions at the junction of the helices cause C_{17} to stack with helix I. This structure is somewhat misleading because multiple divalent metal ions are required in order for the RNA to fold and carry out catalysis. At least one of these metal ions is intimately involved in the catalytic process. The precise number of metal ions that are required for folding and/or catalysis is uncertain. Moreover, the two structures that have

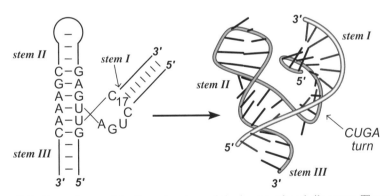

Figure 2-7. Secondary and tertiary structure of the hammerhead ribozyme. The capital letters refer to the bases of the nucleic acids, and the lines show hydrogen bonding interactions between the bases. The cleavage site is indicated by the arrow. Reprinted with permission from D. A. Doherty and J. A. Doudna, *Annu. Rev. Biochem.* **69**, 597 (2000). © 2000 by Annual Reviews. www.annualreviews.org.

been determined are necessarily noncleavable variants: one has a DNA substrate strand and the other a 2'-O-methyl group at the cleavage site. The two structures are not identical, and not entirely consistent with mutagenesis studies that place specific groups at the active catalytic site. This is undoubtedly a reflection of the flexibility of the structure. Nevertheless, knowledge of ribozyme structure is a requisite for understanding their mechanisms of action and for engineering ribozymes for enhanced catalysis and function.

PROTEIN STRUCTURE

Hundreds of protein structures are known, with a great deal of variation in structure, as well as similarities, among them. In this section, some of the common motifs will be discussed, as well as a few examples. Fortunately, the coordinates for protein structures can be found in a single database, the Protein Data Bank, which can be accessed via the internet. Free software also is available on the internet for viewing and manipulating structures with known coordinates, for example, Kinemage (7) and RasMol (8). This software also can be used for nucleic acid structures. Interested readers should explore this software and database, as their use is very important for understanding biological mechanisms on a molecular basis.

The linear sequence in which amino acids are arranged in a protein is termed the *primary* structure. The two most common long-range ordered structures in proteins are the α-helix and the β-sheet. Ordered structures within a protein are called *secondary* structures. Both structures are stabilized by hydrogen bonds between the NH and CO groups of the main polypeptide chain. In the case of the α-helix, depicted in Figure 2-8 (see color plates), a coiled rod-like structure is stabilized by hydrogen bonds between residues that are four amino acids apart. In principle, the screw sense of the helix can be right-handed or left-handed. However, the right-handed helices are much more stable because they avoid steric hindrance between the side chains and backbone. Helices with a different pitch, that is, stabilized by hydrogen bonding between residues other than four apart are found, but the α-helix is the most common helical element found. The helical content of proteins ranges from almost 100% to very little. Helices are seldom more than about 50 Å long, but multiple helices can intertwine to give extended structures over 1000 Å long. A specific example is the interaction of myosin and tropomyosin in muscle.

Beta sheets also are stabilized by NH-CO hydrogen bonds, but in this case the hydrogen bonds are between adjacent chains, as depicted in Figure 2-9 (see color plates). The chains can be either in the same direction (parallel β-sheets) or in opposite directions (antiparallel β-sheets). These sheets can be relatively flat or twisted in protein structures. In schematic diagrams of proteins, α-helices are often represented as coiled ribbons and β-sheets as broad arrows pointing in the direction of the carboxyl terminus of the polypeptide chain (9).

Most proteins are compact globular structures, consisting of collections of α-helices and β-sheets. Obviously if the structure is to be compact, it is necessary for the polypeptide chain to reverse itself. Many of these reversals are accomplished with a common structural element, a reverse turn or β-hairpin bend, that alters the hydrogen bonding pattern. In other cases, a more elaborate loop structure is used. These loops can be quite large. Although they do not have regular periodic structures, analogous to α-helices and β-sheets, they are still very rigid and well-defined structures.

The first protein whose structure was determined by x-ray crystallography to atomic resolution was myoglobin (10). Myoglobin serves as the oxygen transporter in muscle. In addition to the protein, it contains a *heme*, a protoporphyrin with a tightly bound iron. The iron is the locus where oxygen is bound. Myoglobin was a particularly fortuitous choice for structure determination. Of course, good crystals could be obtained. In addition, myoglobin is very compact and consists primarily of α-helices. Its structure is shown in Figure 2-10 (see color plates). The α-helices form a relatively compact electron density so that the polypeptide could be easily traced at low resolution. The structure has several turns that are necessary to maintain the compact structure. Overall the myoglobin molecule is contained within a rectangular box roughly $45 \times 35 \times 25$ Å. The overall trajectory of the polypeptide chain, that is the folding of the secondary structure, is called the *tertiary* structure.

For water-soluble proteins, including myoglobin, the interior is primarily hydrophobic or nonpolar amino acids such as leucine, pheylalanine, etc. Amino acids that have ionizable groups, such as glutamic acid, lysine, etc., are normally on the exterior of the protein. If a protein exists in a membrane, a hydrophobic environment, the situation is often reversed, with hydrophilic residues on the inside and nonpolar residues on the outside. In some cases, however, the structure is hydrophobic on both the "inside" and "outside." Understanding how polypeptides fold into their native structures is an important subject under intense investigation. In an ideal world, the structure of a protein should be completely predictable from knowledge of its primary structure, that is, its amino acid sequence. However, we have not yet reached this goal.

As a second example of a protein structure, the structure of ribonuclease A is depicted in Figure 2-11 (see color plates) (11). This protein is quite compact and contains both α-helix and β-sheet modules with significant loops and connecting structures. In addition, four disulfide linkages are present. This enzyme is kidney shaped with the catalytic site tucked into the center of the kidney. The active site is shown by the placement of an inhibitor and two histidine residues that participate in the catalytic reaction. This structure proved much more elusive than myoglobin since a larger variety of structural elements are present.

Not all proteins are as compact as myoglobin and ribonuclease. In some cases, a single polypeptide chain may fold into two or more separate structural domains that are linked to each other, and many proteins exist as oligomers of polypeptide chains in their biologically active form. The arrangement of oligomers within a protein is called the *quaternary* structure. The first well-studied example of the latter structure was hemoglobin. Hemoglobin is a tetramer of polypeptide chains. Two different types of polypeptides are present: they are very similar but not identical. The typical structure of hemoglobin is $\alpha_2\beta_2$ with α and β designating the two types of quite similar polypeptide chains. Hemoglobin, of course, is used to transport oxygen in blood. In a sense it is four myoglobins in a single molecule. Four porphyrins and irons are present, each complex associated with a single polypeptide chain. However, the interactions between the polypeptide chains are extremely important in the function of hemoglobin and influence how oxygen is released and taken up. In brief, the uptake and release are highly cooperative so that they occur over a very narrow range of oxygen concentration (cf. 12).

Hemoglobin is known to exist in two distinct conformations that differ primarily in the subunit interactions and the details of the porphyrin binding site. The structures of both conformations of hemoglobin are shown in Figure 2-12 (see color plates). One of the conformations binds oxygen much better than the other, and it is the switching between these conformations that is primarily responsible for the cooperative uptake and release of oxygen. A complete discussion of hemoglobin structure is beyond the scope of this presentation, which is focused on illustrating some of the major features of protein structures.

ENZYME CATALYSIS

As final illustration of the application of x-ray crystallography to biology, we consider the enzyme DNA polymerase. Elucidating the mechanism by which enzymes catalyze physiological reactions has been a long-standing goal of biochemistry. X-ray crystallography has been used to probe many enzyme mechanisms. The choice of DNA polymerase is arbitrary, but it clearly is an enzyme at the core of biology and requires the knowledge of both protein and nucleic acid structures. DNA polymerase is the enzyme responsible for replication of new DNA. The reaction proceeds by adding one nucleotide at a time to a growing polynucleotide chain. This addition can only occur in the presence of a *DNA template* that directs which nucleotide is to be added by forming the correct hydrogen bonds between the incoming nucleotide and template.

The molecular structures of a wide variety of DNA polymerases are known (cf. 13). The general structure of the catalytic site is similar in all cases and has been described as a right hand with three domains, fingers, a thumb, and a palm. The structure of a catalytic fragment of the thermostable *Bacillus stearothermophilus* enzyme is shown in Figure 2-13 (see color plates) (14, 15). The fingers and thumb wrap around the DNA and hold it in position for the catalytic reaction that occurs in the palm. To initiate the reactions, a primer DNA strand is required which has a free 3′-hydroxyl group on a nucleotide already paired to the template. The enzyme was crystallized with primer templates, which are included in Figure 2-13. The reaction occurs via a nucleophilic attack of this hydroxyl group on the α-phosphate of the incoming nucleotide, assisted by protein side chains on the protein and two metal ions, usually Mg^{2+}. We will not be concerned with the details of the chemical reaction.

The DNA interacts with the protein through a very extensive network of noncovalent interactions that include hydrogen bonding, electrostatics, and direct contacts. More than 40 amino acid residues are involved that are highly conserved in all DNA polymerases. The interactions occur with the minor groove of the DNA substrate and require significant unwinding of the DNA. The major groove does not appear to interact significantly with the protein and is exposed to the solvent.

The synthesis of DNA is a highly processive reaction, that is, many nucleotides are added to the growing chain without the enzyme and growing chain separating. In the case of the polymerase under discussion, more than 100 nucleotides are added before dissociation occurs. This requires the DNA chain to be translocated through the catalytic site. This translocation has been observed to occur in the crystals by soaking the crystals with the appropriate nucleotides and determining the structures of the products of the reactions. Remarkably, the crystals are quite active, with up to six nucleotides being added and a translocation of the DNA chain of 20 Å.

The enzyme exists in at least two conformations. The initial binding of the free nucleotide occurs in the open conformation, but the catalytic step is proposed to occur in a closed conformation in which the fingers and thumb clamp onto the DNA and close around the substrates. The fidelity of DNA polymerase is remarkable, with only approximately 1 error per 10^5 nucleotides incorporated (16). This is accomplished primarily by the very specific hydrogen bonds occurring between the template and the incoming nucleotide, but interactions with the minor groove also are important. The conformational change accompanying binding of the nucleotide to be added probably also plays a role in the fidelity by its sensitivity to the overall shapes of the reactants. Also the template strand is postulated to move from a "preinsertion" site to an "insertion" site. The acceptor template base that interacts with the incoming nucleotide occupies the preinsertion site in the open conformation and the insertion site in the closed conformation where interaction with the incoming nucleotide occurs. Thus, the addition of each nucleotide is accompanied by a series of conformational changes that translocate the template into position for the next step in the reaction. A movie of this process is available on the internet (15).

Even the high fidelity of the polymerase reaction is not sufficient for the reliable duplication of genes that is required in biological systems. Consequently, the enzyme has a built-in proofreading mechanism, an exonuclease enzyme activity that is located 35 Å or more from the polymerase catalytic site. If an incorrect nucleotide is incorporated, the match within the catalytic site is not perfect, and the polymerase reaction stalls. This brief pause is sufficiently long so that the offending nucleotide base can migrate to the exonuclease site and be eliminated. Unfortunately, the proofreading process itself is not perfect so that every 20 incorporations or so a correct base is chopped off. This wasteful process, however, increases the fidelity of replication by a factor of about 1000.

This very brief discussion of DNA polymerase indicates the great insight into biological reactions that knowledge of macromolecular structures can bring. If the crystals are biologically active, multiple structures with various substrates and inhibitors can provide a detailed mechanistic proposal. This paves the way for additional experiments that will shed even more light on the molecular mechanism.

REFERENCES

1. I. Tinoco Jr., K. Sauer, J. C. Wang, and J. D. Puglisi, *Physical Chemistry: Principles and Applications in Biological Sciences*, 4th edition, Prentice Hall, Englewood, NJ, 2002, pp. 667–711.

2. A. McPherson, *Introduction to Macromolecular Crystallography*, Wiley-Liss, Hoboken, NJ, 2003.

3. K. Moffat, *Chem. Rev.* **101**, 1569 (2001).

4. A. Rich and S. H. Kim, *Sci. Amer.* **238**, 52 (1978).

5. E. A. Doherty and J. A. Doudna, *Annu. Rev. Biochem.* **69**, 597 (2000).

6. W. G. Scott, J. T. Finch, and A. Klug, *Cell* **81**, 991 (1995).

7. D. C. Richardson and J. S. Richardson, *Trends Biochem. Sci.* **19**, 135 (1994).

8. R. A. Sayle and E. J. Milner-White, *Trends Biochem. Sci.* **20**, 374 (1995).

9. J. S. Richardson, D. C. Richardson, N. B. Tweedy, K. M. Gernert, T. P. Quinn, M. H. Hecht, B. W. Erickson, Y. Yan, R. D. McClain, M. E. Donlan, and M. C. Suries, *Biophys. J.* **63**, 1186 (1992).

10. C. L. Nobbs, H. C. Watson, and J. C. Kendrew, *Nature* **209**, 339 (1966).

11. G. Kartha, J. Bello, and D. Harker, *Nature* **213**, 862 (1967).

12. M. F. Perutz, A. J. Wilkinson, M. Paoli, and G. G. Dodson, *Annu. Rev. Biophys. Biomolec. Struct.* **27**, 1 (1998).

13. C. A. Brautigam and T. A. Steitz, *Curr. Opin. Struct. Biol.* **8**, 54 (1998).

14. J. R. Kiefer, C. Mao, J. C. Braman, and L. S. Beese, *Nature* **391**, 304 (1998).

15. S. J. Johnson, J. S. Taylor, and L. S. Beese, *Proc. Natl. Acad. Sci. USA* **100**, 3895 (2003).

16. T. A. Kunkel and K. Bebenek, *Annu. Rev. Biochem.* **69**, 497 (2000).

PROBLEMS

2.1. Assume a lattice of atoms equidistant from each other in all directions (a cubic lattice) with a distance between atoms of 2.86 Å. If a crystal of this material is irradiated with x rays with a wavelength of 0.585 Å, at what angles are Bragg reflections seen for the planes that are 2.86 Å apart? (More than one set of reflections will be seen, but we will not deal with that complexity here.) If the x rays have a wavelength of 6.00 Å, what would be observed?

2.2. The bond energy for a C–C bond is about 340 kJ/mole, 600 kJ/mole for C=C, and 400 kJ/mole for C–H. Calculate the energy/mole associated with x-ray radiation with wavelengths of 0.585 and 1.54 Å. What does this imply about the stability of biomolecules in an x-ray beam?

2.3. When a molybdenum target is used as an x-ray source, x rays of wavelength 0.710 Å are produced. For a particular crystal, a reflection was observed at 4° 48′. What are the possible distances between Bragg planes? The same reflection found when a copper target is used as an x-ray source is found at 10° 27′. What is the wavelength of the x ray produced by the copper target?

2.4. A crystal structure consists of identical units called unit cells. The unit cells are arranged in order, much like a brick wall, to form the crystal. The unit cell contains multiple atoms, and often multiple molecules. Consider one particular plane of a unit cell that extends indefinitely in the x direction. The following structure factors, F_i, were obtained for the planes (Data from P. W. Atkins, *Physical Chemistry*, 3rd ed. W. H. Freeman, New York, 1986, p. 561):

Plane:	0	1	2	3	4	5	6	7	8	9	10	11	12	13	14	15
F_i	16	–10	2	–1	7	–10	8	–3	2	–3	6	–5	3	–2	2	–3

Assume that the electron density, $\rho(x)$, can be calculated from the structure factors by the relationship

$$\rho(x) = F_0 + 2\sum_{i>0} F_j \cos(2\pi i x)$$

Evaluate this sum for x = 0, 0.1, 0.2, 0.3, 0.4, 0.5, 0.6, 0.7, 0.8, 0.9, and 1.0, and construct a plot of $\rho(x)$ versus x. (This is best done with a computer program.) The maxima in the plot correspond to positions of atoms along the x axis. This exercise should give you a good idea of how electron density maps are obtained. The equation for $\rho(x)$ is an example of a Fourier transformation.

2.5. Go to the internet address www.proteinexplorer.org and do the 1-hour tour. This will allow you to make molecular drawings of proteins and nucleic acids whose structures are known.

CHAPTER 3

ELECTRONIC SPECTRA

INTRODUCTION

The most common type of spectroscopy involves light in the visible and ultraviolet region of the spectrum interacting with molecules. This interaction causes electrons to shift between their allowed energy levels. If light shines on a colored sample, some of the radiation is absorbed by the sample. This absorption can heat the sample, it can cause photons of the same or lower energy to be emitted, or photochemical reactions can occur. We will not be concerned with the latter phenomenon. Of course, in all cases energy must be conserved, that is, the total energy of all of the three processes must be equal to the energy of the photons that are absorbed.

The absorption of light is due to the interaction of the oscillating electromagnetic field of the radiation with the charged particles in the molecule. If the electromagnetic force is sufficiently large, the electrons in the molecule are rearranged, that is, shifted to higher energy levels. The absorption process is extremely fast, occurring in about 10^{-15} s. The excited state equilibrates with its surrounding and returns to the ground electronic state either by the production of heat or the emission of radiation with an energy equal to or less than that of the absorbed radiation. The latter process is called *fluorescence* and typically occurs in nanoseconds. The absorption process is shown schematically in Figure 3-1; a more exact description is discussed later (Fig. 3-11).

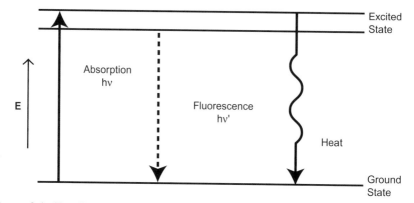

Figure 3-1. Simplified schematic diagram of electron excitation by absorbance of radiation. An electron is excited to a higher energy level by radiation of frequency υ. It returns to its ground state through fluorescence at a frequency υ' and/or by dissipation of heat. A more complete diagram is presented in Figure 3-11.

The details of light absorption by a molecule can be determined directly from quantum mechanical calculations. A transition of electrons from one energy level to another occurs when the wave functions for the two electronic states in question are electrically coupled. The detailed calculation tells us what transitions between the energy levels are allowed, for example, from energy level 1 to energy level 2. Electrons can only move between certain energy states; not all transitions are allowed. These *selection rules* are important because they allow us to predict what wavelengths of light will be absorbed and emitted. Although this facet of spectroscopy will not be of great concern here, it is important to remember that transitions of electrons between energy levels are governed by a set of rules. Exceptions to selection rules sometimes occur, but with a much lower probability than the favored event. This is primarily due to the fact that the potential energy functions used in the quantum mechanical calculations are not perfect.

ABSORPTION SPECTRA

From a practical standpoint, the most important aspect of the quantum mechanical calculation is the determination of how much light is absorbed by the sample. This is embodied in the Beer-Lambert law, which gives the relationship between the light intensity entering the solution, I_0, and the light intensity leaving the solution, I:

$$\log(I_0/I) = A = \varepsilon c l \tag{3-1}$$

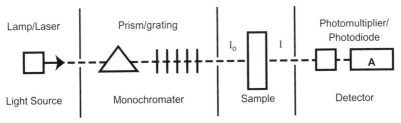

Figure 3-2. Schematic representation of an absorption spectrophotometer. Common light sources are tungsten and arc lamps and lasers. Monochromatic light of intensity I_0 is obtained with a prism or diffraction grating. The detection of the transmitted light of intensity I is with a photomultiplier or photodiode, with the electronic signal converted to the absorbance, A.

Here A is defined as the absorbance, ε is the molar absorbtivity or extinction coefficient, c is the concentration of the absorbing material, and l is the thickness of the sample through which the light passes. In visible-ultraviolet spectrophotometers, a solution of the sample is put into a cuvette of calibrated dimensions, and visible or ultraviolet light is passed through the sample. The intensity of the light before and after passing through the solution is determined through the use of a variety of detectors, such as photomultipliers and photodiodes. A schematic diagram of a typical spectrophotometer is shown in Figure 3-2. The extinction coefficient is a different constant for each wavelength and is characteristic of the molecule under investigation. In principle, it can be calculated through quantum mechanics, but more practically it can be determined experimentally. A spectrophotometer also must account for light lost by reflection, absorption by the cuvette, and various other factors.

Deviations from the Beer-Lambert law can be observed in the laboratory. For example, the light might not be monochromatic, the sample might be inhomogeneous, light scattering could occur, the photodetectors might exhibit nonlinear behavior, etc. These are instrumental and sample artifacts. However, apparent deviations also occur that are due to the molecular properties of the sample: for example, sample aggregation or complex formation as the concentration is changed. In this case, the deviations can be used to study these properties. Experimental studies must carefully verify the validity of the Beer-Lambert law over the concentration range being studied.

A very important property of light absorbing solutions is that if multiple absorbing species are present, the total absorbance is simply the sum of the absorbance of the individual species. For example, for species X and Y, the absorbance is

$$A_\lambda = A_\lambda^X + A_\lambda^Y = \varepsilon_\lambda^X l(X) + \varepsilon_\lambda^Y l(Y) \tag{3-2}$$

If measurements are made at two different wavelengths where the extinction coefficients are sufficiently different, it is possible to determine the concentrations of the individual species if the extinction coefficients are known. Equation 3-2 generates two equations with different numerical values of the extinction coefficients at the two wavelengths. These two equations can be solved simultaneously to give the two concentrations. This is a very common method for determining the concentrations of individual species in solutions. In fact, this approach can be extended to more complicated situations. For example, if three species are present, absorbance measurements at three different wavelengths can be used to determine the concentrations of the individual components.

The wavelength at which two components have exactly the same extinction coefficient is called the *isobestic* wavelength. In this case, it is obvious that measurement of the absorbance determines the total concentration of the two components:

$$A_{isobestic} = \varepsilon_{isobestic} l[(X)+(Y)] \tag{3-3}$$

Although one or more isobestic wavelengths may be found for any pair of components, this is not necessarily the case, as the two components may not have significant absorbance in the same wavelength region. A useful rule to remember is that if two or more isobestic wavelengths are seen at a series of different concentrations and/or varying conditions such as pH, then only two components are present. (This is not entirely rigorous, but the probability of this not being the case is exceedingly small.) This is especially useful when examining samples in which the individual extinction coefficients are not known.

ULTRAVIOLET SPECTRA OF PROTEINS

Proteins that do not contain strongly absorbing extrinsic cofactors such as hemes have a very characteristic ultraviolet spectrum, whereas they are essentially transparent in the visible region of the spectrum. The spectrum of serum albumin is shown in Figure 3-3 as a plot of the absorbance versus the wavelength. It has two absorption maxima, one at about 280 nm, the other, much stronger absorption, occurs around 190—210 nm. The absorption at 280 nm is frequently used to measure protein concentrations. Although the absorption is much stronger at 200 nm, it is harder to make measurements at this wavelength so absorption at 200 nm is not routinely used as an analytical tool. The ultraviolet spectrum of a protein is sufficiently characteristic to use as a preliminary indication that a protein is present.

The spectra of the aromatic amino acids phenylalanine, tyrosine, and tryptophan are shown in Figure 3-4. It can be readily seen that electronic transitions in tyrosine and tryptophan are primarily responsible for the absorption

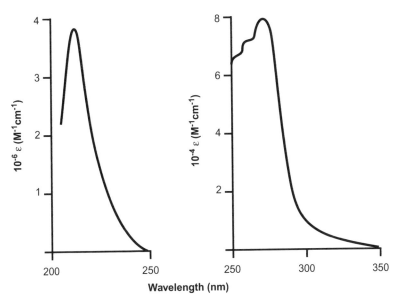

Figure 3-3. Ultraviolet absorption spectrum of bovine serum albumin at pH 7.0. The extinction coefficient is plotted versus the wavelength in two panels because of the large difference in the ordinate scale as the wavelength changes. Data from E. Yang in I. Tinoco, K. Sauer, J. C. Wang, and J. D. Puglisi, *Physical Chemistry,* 4th edition, Prentice Hall, Englewood Cliffs, NJ. 2002, p. 548.

peak at around 280 nm. It can also be seen that all three amino acids absorb strongly at shorter wavelengths. Measurements of the spectra of polypeptides not containing aromatic amino acids show strong absorbance at around 192 nm due to electronic transitions associated with the peptide bond. Thus, the absorbance maximum in the 190—210 nm range is due to both aromatic amino acids and peptide bonds. The absorption due to side chains of other amino acids is typically at least an order of magnitude less in this wavelength region.

The spectra of aromatic amino acids and peptide bonds in proteins are significantly influenced by their local environments. This is because the spectra are fundamentally due to electrical phenomena so that structural features that determine the local charge distribution and the local dielectric constant strongly influence the spectra. Thus, the spectra of peptide bonds and aromatic amino acids buried inside the protein are somewhat different than those on the exterior of the protein. Hydrogen bonding and ion pair interactions also influence the spectra. This environmental sensitivity of protein spectra can be used to obtain information about the structure of proteins. For example, the amount of α-helix, β-sheet, and random coil can be estimated from the ultraviolet spectrum (1). A denatured protein usually has a somewhat different ultraviolet spectrum than the native protein so that protein folding reactions

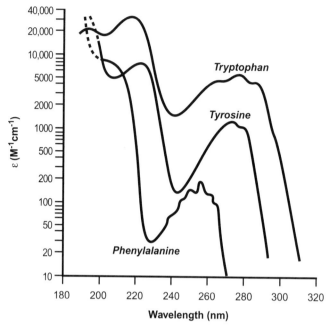

Figure 3-4. Absorption spectra of phenylalanine, tyrosine, and tryptophan at pH 6.0. Note the logarithmic scale on the ordinate. Adapted from S. Malik in D. B. Wetlaufer, Ultraviolet Spectra of Proteins and Amino Acids, *Adv. Prot. Chem.* **17**, 303 (1962). © 1962, with permission from Elsevier.

can be monitored by measuring spectral changes. However, the change in the ultraviolet spectrum is usually rather small so that other methods have proved more useful for assessing secondary structure and monitoring protein folding (e.g., circular dichroism, which is discussed later).

NUCLEIC ACID SPECTRA

In contrast to proteins, all of the common nucleotide bases have quite similar absorption spectra, with an absorption peak at around 260 nm. The spectra of adenine, thymine, cytosine, uracil, and guanine are shown in Figure 3-5. Nucleic acid solutions also are essentially transparent in the visible region of the spectrum. The intense absorption at 260 nm is frequently used as an indication of the presence of nucleic acids and can be used to determine their concentrations by use of the Beer-Lambert law.

Again, the local environment of individual nucleotide bases can influence the spectrum. Generally, nucleic acids have a lower absorbance than the sum of the absorbance of the individual nucleotides. This phenomenon is called

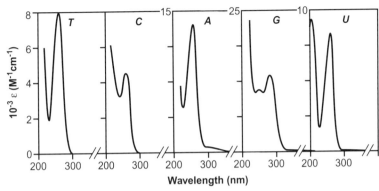

Figure 3-5. Absorption spectra in water of the four bases commonly found in nucleic acids: thymine (T), cytosine (C), adenine (A), guanine (G), and uracil (U). Data from H. Du, R. A. Fuh, J. Li, A. Corkan, and J. S. Lindsey, *Photochem. Photobiol.* **68**, 141 (1998), [http://omlc.ogi.edu/spectra/PhotochemCAD/].

hypochromicity and is due to the "stacking" or lining up of parallel planes of the bases. An increase in absorption is called *hyperchromicity*. These phenomena are very useful for determining whether nucleic acids are structured. For example, native (double-helix) DNA displays significant hypochromicity, and if the DNA is unwound, the absorbance at 260 nm increases.

Although the stable structure of DNA is the well-known double helix, the double-helix structure is lost as the temperature increases, and at a sufficiently high temperature, the two strands of DNA separate. This loss of structure can be monitored by the increase in absorbance at 260 nm, as shown schematically in Figure 3-6. Obviously, the more stable the DNA, the higher the temperature that is needed to break down the structure so that doing a temperature "melt" of DNA is a useful tool for determining the stability of a specific DNA. With this technique, for example, it is easy to show that G-C base pairs are more stable than A-T pairs. By studying the stability of a series of nucleic acid oligomers, it is possible to predict the stability of DNA and, less reliably, RNA structures (cf. 2). Also, when DNA is synthesized, the double-helix structure must be broken in order to provide a single-strand template for the polymerization reaction. This unwinding of DNA *in vivo* is done by helicase enzymes. Obviously, monitoring the DNA structure is important for understanding DNA synthesis, in general, and the mechanism of action of helicases, in particular. This subject is considered further later in this chapter.

PROSTHETIC GROUPS

Some proteins contain tightly bound organic molecules and/or metal ions. These are usually referred to as *prosthetic groups* and are essential for the

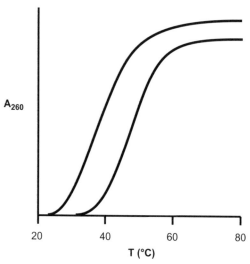

Figure 3-6. Schematic representation of DNA melting for two different DNA. The absorbance at 260 nm increases as the temperature is raised and the double-stranded structure breaks down. The temperature at which the melting occurs increases as the amount of G-C base pairs increases.

biological activity of the macromolecules. The spectra of these prosthetic groups are often in the visible region of the spectrum, easily separable from the absorption due to the protein, and can be monitored to follow the reactions undergone by the protein.

One of the most common prosthetic groups is *heme*, the structure of which is shown in Figure 3-7. The heme binds iron tightly, either as Fe^{2+} or Fe^{3+}. Probably the most well-known heme protein is hemoglobin, mentioned in Chapter 2. Hemoglobin has a molecular weight of about 64,000. It has four polypeptide chains and contains four hemes. It is, of course, responsible for the transport of oxygen in the body. The structure of the protein is well known (Fig. 2-12, see color plates) and its function has been extensively studied (cf. 3). For the purposes of this discussion, we consider only the visible spectra of the molecule. The active molecule contains Fe^{2+}: when oxygen is bound, it is coordinated directly to the iron. In the absence of oxygen, this iron coordination site is occupied by water. The visible spectra of oxy- and deoxy-hemoglobin are quite distinct, as shown in Figure 3-8. This is because the binding of oxygen alters the environment of the iron/heme, thereby altering the electronic energy levels. Thus, the visible spectrum can be used to monitor the binding of oxygen. Oxy-hemoglobin is responsible for the bright red color of oxygenated blood. The ferrous iron can be oxidized to the ferric form, which also alters the spectrum of the molecule. This form of hemoglobin is not involved in oxygen transport. The molecular mechanism for the binding of oxygen to hemoglobin is now understood in molecular detail. It is a highly cooperative process with

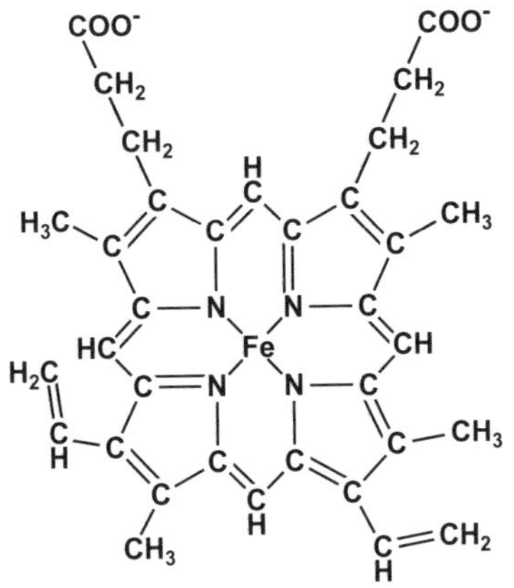

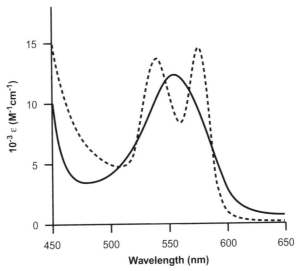

Figure 3-7. Structure of the heme found in hemoglobin. The Fe is in the oxidation state +2 when oxygen is bound.

Figure 3-8. Visible absorption spectra of oxy-hemoglobin (*dashed line*) and deoxyhemoglobin (*solid line*). These two forms of hemoglobin can be readily distinguished with absorption spectroscopy. Adapted from L. Stryer, *Biochemistry*, 2nd edition, W. H. Freeman, San Francisco, 1981, p. 52. © 1976, 1981 by Lubert Stryer. Used with permission of W. H. Freeman and Company.

alterations in the position of the iron in the molecule and significant conformational changes within the protein accompanying binding.

Cytochromes also contain hemes. For these molecules the oxidation-reduction of iron is crucial to their physiological function. The progress of the oxidation-reduction reactions can be readily monitored through changes in the visible spectrum.

Flavins constitute another type of common prosthetic group. Flavin mononucleotide (FMN) and flavin adenine dinucleotide (FAD) are important constituents in many enzymes that participate in redox reactions. Their structures are shown in Figure 3-9. These flavin prosthetic groups can exist in three oxidation states: oxidized, semiquinone radical, and reduced. (A radical is an organic molecule with an unpaired electron.) All three of these oxidation states have different absorption peaks in the visible. The three oxidation states and their absorption maxima are included in Figure 3-9. The semiquinone can also lose a hydrogen to give a diradical (two unpaired electrons) with yet a different absorption maximum. The pK for the loss of this hydrogen is about 8.4. Flavin enzymes can undergo both one- and two-electron oxidation-reduction reactions and can be readily reoxidized by oxygen. The visible spectrum serves as a useful indicator of the oxidation state of the flavin in an enzyme. It can also be used to monitor the reaction progress of the enzymatic reactions.

Why do these prosthetic groups have electronic transitions at longer wavelengths than proteins? This can be readily understood in molecular and quantum mechanical terms. Both hemes and flavins have highly aromatic structures, that is, extensive conjugation occurs. As a result, the electrons are delocalized around the ring structures (π electrons). If the electrons are considered as "particles in a box," as discussed in Chapter 1, the size of the box is increased relative to less conjugated structures such as tyrosine, phenylalanine, and tryptophan. An increase in the size of the box increases the wavelengths that characterize the electronic transitions (Eq. 1-11).

DIFFERENCE SPECTROSCOPY

Because the absorption of light by proteins and nucleic acids depends on the local environment of the chromophores, changes in light absorption often occur when small molecules bind to proteins or nucleic acids and when macromolecules interact. As an example, we consider the binding of nucleotides to the enzyme ribonuclease A. As the name implies, ribonuclease A hydrolyzes RNA. It is a small protein with a molecular weight of approximately 14,000 and contains tyrosine, tryptophan, and phenylalanine so that it has a typical protein ultraviolet spectrum. Its structure is shown in Figure 2-11 (see color plates). It binds nucleic acids and has a great preference for cleavage adjacent to a pyrimidine. When nucleotides bind to the enzyme, the ultraviolet spectrum is changed. However, the changes are very small so that a technique called difference spectroscopy must be used.

With difference spectroscopy, a double-beam spectrophotometer is used. Each beam contains two cuvettes. For one of the beams, one cuvette contains the nucleotide and the other the protein. In the other beam, one of the cuvettes

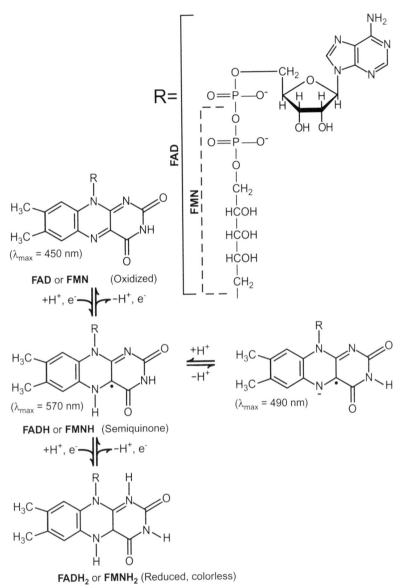

Figure 3-9. Oxidation states of FAD and FMN with the absorption maxima indicated. These oxidation states are important for the biological function of these molecules and can be readily distinguished with absorption spectroscopy. The pK of the semiquinone is about 8.4.

contains a mixture of the nucleotide and protein and the other contains buffer. The concentration of nucleotide and of enzyme is exactly the same in each beam. Consequently, any difference in absorbance that is measured must be due to the interaction of the nucleotide and the protein. The absorbance in the beam containing the mixture of enzyme and nucleotide (assuming a path length of 1 cm) is

$$A_1 = \varepsilon_{EL}(EL) + \varepsilon_E(E) + \varepsilon_L(L)$$
$$= \varepsilon_{EL}(EL) + \varepsilon_E[(E_0) - (EL)] + \varepsilon_L[(L_0) - (EL)] \qquad (3\text{-}4)$$

where the ε's are the extinction coefficients for the enzyme, E, the ligand, L, and the enzyme-ligand complex, EL. The study of ligand binding showed that only one nucleotide binds per enzyme molecule. The absorbance in the other beam is given by

$$A_2 = \varepsilon_E(E_0) + \varepsilon_L(L_0) \qquad (3\text{-}5)$$

Subtraction of Eq. 3-5 from Eq. 3-4 gives the difference absorbance

$$\Delta A = (\varepsilon_{EL} - \varepsilon_E - \varepsilon_L)(EL) = \Delta\varepsilon(EL) \qquad (3\text{-}6)$$

Thus, measuring the difference absorbance is a direct measure of the concentration of the enzyme-nucleotide complex.

Titration of the enzyme with the nucleotide and extrapolation to a concentration where all of the ligand is bound to the enzyme permits determination of the difference extinction coefficient, $\Delta\varepsilon$. The difference extinction coefficient for the binding of 2'-cytidine monophosphate to ribonuclease is shown in Figure 3-10 at different wavelengths. Note that significant changes are found at wavelengths of 260 and 280 nm, indicating that the extinction coefficients of both the enzyme and nucleotide are altered in the complex. Note also the isobestic point that is found as the pH is varied, demonstrating the presence of only two absorbing species, bound and unbound. Since the concentration of all species can be determined during the titration, the binding constant can be calculated and was found to be $1.8 \times 10^5 \, M^{-1}$ at pH 5.5. An extensive study carried out with various ligands over a range of pH provided valuable information about the molecular details of the interaction between the ligands and enzyme (4).

X-RAY ABSORPTION SPECTROSCOPY

X rays can also be absorbed by materials, and the Beer-Lambert law (Eq. 3-1) is applicable. The energy associated with an x ray of 1 Å wavelength is about $10^6 \, kJ/mole$ ($E = hc/\lambda$). This energy is sufficiently large that the

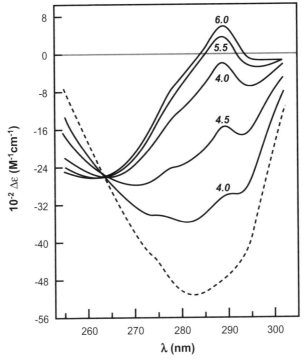

Figure 3-10. Difference absorption spectra of ribonuclease A-2′-CMP as a function of pH. The pH is indicated next to each difference spectrum. The dashed line is the calculated difference spectra for the protonated ring species of 2′-CMP. Reprinted in part with permission from D. G. Anderson, G. G. Hammes, and F. G. Walz Jr., *Biochemistry* **7**, 1637 (1968). © 1968 by American Chemical Society.

absorption of x rays by atoms results in electrons being ejected from inner orbits (1s, 2s, 2p). The x-ray absorption spectra of atoms consist of sharp peaks, the peaks occurring at different wavelengths for every element in the periodic table. If the atom is in a molecule, the ejected electron interacts with the environment, resulting in small oscillations around the main peak in the absorption spectra due to back scattering. Analysis of this fine structure provides information about the nature of the atoms surrounding the atom that has absorbed the x rays. This is particularly useful for probing the environment around heavy metals in biological structures.

FLUORESCENCE AND PHOSPHORESCENCE

Fluorescence is a very common phenomenon in biology and elsewhere. As previously discussed, fluorescence is the emission of light associated with electrons moving from an excited state to the ground state. Fluorescence and phosphorescence are often more useful tools for studying biological process than absorbance. They can be considerably more sensitive than absorbance so that much lower concentrations can be detected. They are also often extremely sensitive to the environment and thus are extremely good probes of the local environment. Some other useful properties will be discussed a bit later.

In order to understand fluorescence and phosphorescence, a somewhat more complex energy diagram such as Figure 3-11 is useful. In this diagram, two electronic energy levels are shown, each with a manifold of vibrational energy levels. The vibrational energy levels are much more closely spaced than electronic energy levels and are due to the vibrations of atoms within a molecule. Different modes of vibration have different energies so that manifolds of quantized energy levels exist. The properties of vibrational energy levels can be calculated through quantum mechanics. Generally, electrons are in their ground electronic energy state at room temperature, and the molecules are also in their lowest vibrational energy level. They can be excited by light to the next electronic energy level, but can be in many different vibrational energy levels in the excited state. After this excitation, the molecule will return

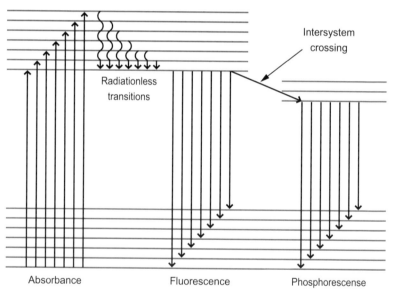

Figure 3-11. Schematic energy level diagram for absorption, fluorescence, and phosphorescence. This is often referred to as a Jablonski diagram. Electrons are excited from a ground state singlet energy level, S_0, to various vibrational states of a higher electronic energy level, S_1. All electrons are paired in singlet states, that is, their spins are in opposite directions. Radiationless decay occurs to the ground vibrational energy level of S_1. As the electrons return to their ground electronic state, fluorescence occurs. In some cases, the excited electrons can move to a triplet state, T_1, in a radiationless transition. Triplet states have two unpaired electrons. The electrons in the triplet state can then return to the ground electronic state with phosphorescence occurring.

to the lowest vibrational energy level of the second electronic energy level very rapidly, in 10^{-13} s or less. This generally occurs through a radiationless transition, that is, through the production of heat. Electrons in the second energy level can then decay to the electronic ground state, again in various vibrational energy levels. The emission produced is fluorescence. The decay to the ground electronic state can also occur without the emission of light, that is, by producing heat.

From Figure 3-11, it is clear that the emitted light must be at a longer wavelength than the absorbed light since the energy change associated with emission can never be greater than the energy change associated with absorption. Also from this diagram it is clear, that the number of photons emitted can never exceed the number of photons absorbed. The efficiency of fluorescence is characterized by the *quantum yield*, Q, which is the fraction of photons absorbed that are eventually emitted:

$$Q = \frac{number\ of\ photons\ emitted}{number\ of\ photons\ absorbed} \tag{3-7}$$

In practice, the quantum yield can be quite close to 1, but for most molecules that fluoresce well, it is usually in the range of 0.3–0.7. Not all molecules that absorb light produce measurable fluorescence; in fact, most do not. Instead, decay to the ground state is radiationless, via heat production, and the quantum yield is essentially zero. The only common amino acid that has a useful fluorescence spectrum is tryptophan, and none of the common nucleic acids display significant fluorescence.

The measurement of fluorescence (and phosphorescence) is more complex than the measurement of absorption since the sample must be excited at a specific wavelength and the fluorescence observed at a different wavelength. Typically, the emission is detected perpendicular to the excitation beam. The emission spectrum is obtained by keeping the excitation wavelength constant and observing the emission over a range of wavelengths. The fluorescence of tryptophan, an amino acid found in many proteins, is shown in Figure 3-12. Also shown is the absorption spectrum. The emission spectrum in the figure was determined by exciting tryptophan with light at the absorption maximum of 275 nm. The excitation spectrum is obtained by observing the emission at a given wavelength and varying the excitation wavelength. The quantum yield is different for each pair of excitation and emission wavelengths. An absolute measurement of fluorescence would require knowledge of the number of photons absorbed at the excitation wavelength and the number of photons emitted at the emission wavelength. Such measurements are very difficult to make so that in practice the fluorescence intensity is usually expressed in relative terms or in comparison with a standard solution where the quantum yields are well known.

A more quantitative description of fluorescence is useful for understanding the experimental measurements. The number of excited molecules is pro-

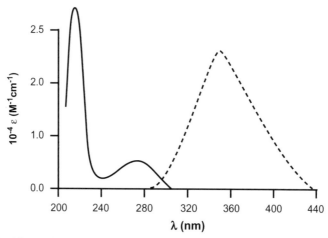

Figure 3-12. Absorption and fluorescence spectrum of tryptophan. The solid line is the extinction coefficient, and the dashed line is the fluorescence emission in arbitrary units with excitation at approximately 275 nm. D. Freifelder, *Physical Biochemistry,* 2nd edition, W. H. Freeman, New York, 1982, p. 539. © 1976, 1982 by W. H. Freeman and Company. Used with permission.

portional to the amount of light absorbed, which can be determined from the Beer-Lambert law:

$$I_t = I_0 e^{-2.3\varepsilon(\lambda)cl} \tag{3-8}$$

Here, I_t is the intensity of the light of incident intensity I_0 after it has passed a distance l through a solution of concentration c with an extinction coefficient of $\varepsilon(\lambda)$ at wavelength λ. Fluorescence measurements are made with solutions of low absorbance, so that the exponential can be expanded to give

$$I_t = I_0[1 - 2.3\varepsilon(\lambda)cl] \tag{3-9}$$

Thus, the concentration of excited molecules is proportional to the light absorbed,

$$I_0 - I_t = 2.3\varepsilon(\lambda)clI_0 \tag{3-10}$$

The fluorescence emitted, F, is the product of the concentration of excited molecules times the quantum yield times the fraction of emitted photons collected by the instrument,

$$F = (\text{constant})Qc \tag{3-11}$$

From the above discussion it is clear that the constant contains several proportionality relationships that are not easily determined, thus necessitating the determination of relative fluorescent intensities rather than absolute intensities. From a practical standpoint, it is important to note that the fluorescence is proportional to the concentration *if* the absorbance of the solution is sufficiently low to permit the expansion of the exponential. Fortunately, this situation is usually quite easy to obtain experimentally.

The discussion thus far is valid for *singlet* electronic states, that is, states in which all of the electron spins are paired. Many complex molecules have electronic states in which two unpaired electrons exist, a *triplet* electronic state. These are at a higher energy state than the ground singlet state and can be at a lower energy than the first excited singlet state. This is shown schematically in Figure 3-11. In principle, the molecule can move from the excited singlet state to the triplet state, and then decay can occur to the ground state via light emission or the production of heat. This is termed *phosphorescence*. As with fluorescence, the emitted light must be at a longer wavelength than the absorbed light.

What makes phosphorescence different than fluorescence, and how can they be distinguished? In the case of fluorescence, quantum mechanics tells us that the transitions between energy states are allowed, and they occur very rapidly, typically in nanoseconds. In the case of phosphorescence, quantum mechanics tells us that the interconversion from a single state to a triplet state is not allowed, that is, it will not occur readily. The result is that phosphorescence generally takes place in milliseconds or longer. Although this qualitative distinction between fluorescence and phosphorescence is usually sufficient to distinguish between the two, the proof of what is occurring requires determination of the magnetic properties of the excited state. Because the triplet state has unpaired electron spins, its energy will be altered in a magnetic field, whereas this will not happen for fluorescence. At room temperature, DNA does not phosphoresce, but in frozen solutions, thymidine will phosphoresce. We will not explicitly consider phosphorescence further, but it should be remembered that it is very similar to fluorescence; moreover, many biological systems display phosphorescence, perhaps most notable are plant and animal life in the ocean.

Figure 3-11 is a considerable simplification of the real situation where many different electronic states are possible so that multiple electronic energy levels must be taken into account. In addition, many different vibrational modes are possible, so multiple manifolds of vibrational energy levels are generally present. However, the principles enunciated are equally valid for these more complex situations.

RecBCD: HELICASE ACTIVITY MONITORED BY FLUORESCENCE

As mentioned above, part of the mechanism of DNA synthesis involves unwinding DNA. RecBCD is a protein from *Escherichia coli* that functions as a helicase (cf. 5). It also has an endo- and exo-nuclease activity, that is, it hydrolyzes DNA. The helicase activity requires the hydrolysis of MgATP to furnish the free energy necessary for the unwinding process. Monitoring absorbance is not a useful assay for following the unwinding process because of the small changes in absorbance accompanying unwinding relative to the total absorbance. However, a fluorescence assay was developed by Roman and Kowalczykowski (5) that provides a continuous assay of the helicase activity. This assay makes use of a protein that binds specifically to single-stranded DNA, SSB. The binding process stabilizes the single-stranded DNA. Fortunately, when SSB binds to single-stranded DNA, the fluorescence of SSB due to tryptophan is quenched. If SSB is put into an assay mixture containing double-stranded DNA, RecBCD, and MgATP, the SSB will bind to the single-stranded DNA as it is produced. The result is a decrease in the tryptophan fluorescence as more and more single-stranded DNA is formed. This change in fluorescence is a direct monitor of the rate of DNA unwinding since the binding of SSB to single-stranded DNA is fast relative to the rate of the unwinding reaction. The apparent turnover number for the enzyme determined by this method ranged from 14 to 56 µM base pairs s^{-1}(µM recBCD)$^{-1}$, depending on the experimental conditions.

The mechanism of action of the recBCD helicase is postulated to proceed by the enzyme binding to DNA and unwinding many base pairs before it dissociates from the DNA. An important question to ask is how much unwinding occurs before the helicase dissociates. When an enzyme catalyzes many events during a single binding, the reaction is called *processive*: the greater the amount of unwinding/binding event, the greater the processivity of the enzyme activity. With the assay described above, it was possible to quantitate the processivity (6). The processivity increases in a hyperbolic manner as the MgATP concentration increases, reaching a limiting value of 32 kilobases at saturating concentrations of MgATP. The apparent dissociation constant for MgATP deduced from the hyperbolic reaction isotherm is about 40 µM. The processivity was found to be quite sensitive to salt concentration. Thus, a substantial amount of DNA is unwound during each event, ultimately resulting in the synthesis of long stretches of DNA by the polymerase enzyme.

This is but one example of the use of fluorescence to develop an assay for an important biological process, with the subsequent determination of mechanistic aspects of the process.

FLUORESCENCE ENERGY TRANSFER: A MOLECULAR RULER

Thus far we have considered only what happens within a single molecule when light is absorbed. However, it is also possible for a molecule in an excited electronic state to transfer the energy above the ground state to another molecule. This phenomenon is termed *Förster energy transfer* (7). This can be readily understood if a simple kinetic scheme is considered. If the energy donor, D, is considered, fluorescence can be described by the following sequence of events:

$$D \xrightarrow{hv} D^* \text{(light absorption)}$$

$$D \xrightarrow{k_f} D + hv' \text{(fluorescence of donor)}$$

$$D^* \xrightarrow{k_r} D \text{(radiationless de-excitation)} \quad (3\text{-}12)$$

The k's are rate constants in this scheme, v is the light frequency, and * designates the excited electronic state. The "natural" fluorescence lifetime is defined as

$$\tau_0 = 1/k_f \quad (3\text{-}13)$$

This lifetime cannot be measured experimentally but is a useful theoretical concept.

The fluorescence lifetime is measured by exciting the molecule with short pulses of light and following the decay of fluorescence. According to Eq. 3-12,

$$-d(D^*)/dt = (k_f + k_r)(D^*) = (D^*)/\tau_D \quad (3\text{-}14)$$

so that the lifetime determined experimentally is

$$\tau_D = 1/(k_f + k_r) \quad (3\text{-}15)$$

Eq. 3-14 can be easily integrated with the boundary conditions that $(D^*) = (D_0^*)$ when $t = 0$ and $(D^*) = 0$ when $t = \infty$. The result is

$$(D^*) = (D_0^*) \exp(-t/\tau_D) \quad (3\text{-}16)$$

Thus, a plot of the natural logarithm of the fluorescence intensity versus time is a straight line with a slope of $-(1/\tau_D)$.

The quantum yield for this sequence of events is simply the fraction of molecules fluorescing after excitation, or

$$Q_D = k_f/(k_f + k_r) = \tau_D/\tau_0 \quad (3\text{-}17)$$

If the frequency of the fluorescence happens to overlap the absorption of a second molecule, the acceptor (A), then some of the energy of de-excitation can be transferred directly to A, which in turn can fluoresce. The scheme in Eq. 3-12 can be expanded to include this possibility:

$$D^* + A \xrightarrow{k_T} D + A^* \text{(energy transfer)}$$

$$A^* \longrightarrow A + hv'' \quad (3\text{-}18)$$

It should be noted that this energy transfer can occur even if A does not fluoresce, as A could return to its ground state without producing light. The quantum yield of D in the presence of A is now given by

$$Q_{DA} = k_f / (k_f + k_r + k_T) \tag{3-19}$$

The efficiency of energy transfer, E, is

$$E = k_T / (k_f + k_r + k_T) = 1 - Q_{DA} / Q_D = 1 - \tau_{DA} / \tau_D \tag{3-20}$$

where τ_{DA} is the fluorescence lifetime in the presence of the acceptor. In this case,

$$-d(D^*)/dt = (k_f + k_r + k_T)(D^*) \tag{3-21}$$

Thus, the efficiency of energy transfer can be measured quite readily.

This efficiency depends very strongly on the physical separation of the acceptor and donor, and therefore can be used to calculate the distance between A and D. The dependence of the efficiency on the distance between A and D, R, is given by

$$E = R_0^6 / (R_0^6 + R^6) \tag{3-22}$$

where

$$R_0 (\text{in nm}) = 8.79 \times 10^{-6} (J\kappa^2 n^{-4} Q_D)^{1/6} \tag{3-23}$$

In Eq. 3-23, J is a measure of the spectral overlap of the fluorescence emission of the donor with the absorption of the acceptor, κ is a measure of the relative orientations of the donor and acceptor, and n is the refractive index. Equation 3-22 was first enunciated by Förster, hence the name Förster energy transfer. Because R_0 can be readily calculated, R can be determined directly from measurements of E. Since E depends on the sixth power of the distance, it is a very sensitive measure of the distance. Typical values of R_0 are 3—6 nm so that distances in the range of 1—10 nm are readily accessible to this method. The measurement of energy transfer has proved very useful for providing structural information about biological systems, especially in cases where the molecular structures are not well known.

APPLICATION OF ENERGY TRANSFER TO BIOLOGICAL SYSTEMS

The use of energy transfer as a molecular ruler in macromolecular systems was elegantly tested by Stryer and Haugland (8). They synthesized a series of proline polymers with a fluorescent donor, dansyl, at one end and an energy acceptor, naphthyl, at the other end. Proline was selected because the polymer is a rigid cylinder whose length can be easily calculated. The number of prolines in the polymer varied from 1 to 12, and R varied from about 1.2 to 4.6 nm. The energy transfer efficiency obeyed Eq. 3-22 exceedingly well, as

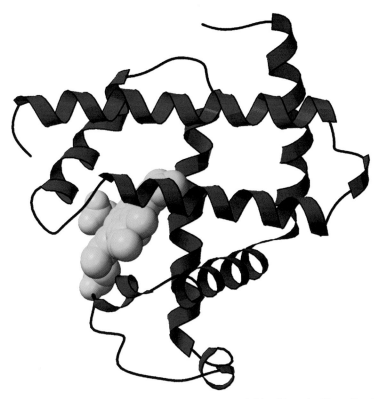

Figure 2-10. Ribbon structure of sperm whale myoglobin (Protein Data Bank entry 1A6M). The large amount of α-helical structure is apparent. The space filling structure in cyan is the heme. Copyright by Professor David C. Richardson. Reprinted with permission. Kinemage graphics, then rendered in Raster3D.

Figure 2-11. Schematic structure of ribonuclease A with uridine vanadate (magenta) bound to the active site (Protein Data Bank entry 1RUV). Both α-helical and β-sheet structure can be seen, as well as structure that does not have a regular array of hydrogen bonds. The disulfide linkages are in yellow, and the green residues are histidines 12 and 119, which are essential for catalytic activity. Copyright by Professor David C. Richardson. Reprinted with permission. Kinemage graphics, then rendered in Raster3D.

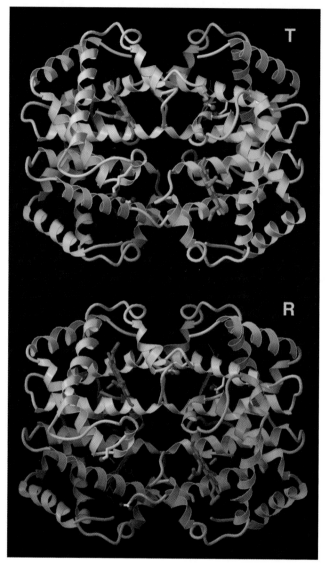

Figure 2-12. Schematic representation of the R (*bottom*, pink) and T (*top*, blue) forms of the hemoglobin $\alpha_2\beta_2$ tetramer. The hemes where the oxygen binds can be seen in the structure. The yellow side chains form salt bonds in the T structure that are broken tin the R. structure. One pair of the α-β subunits also rotates with respect to the other by about 15° in the interconversion of R and T forms. Copyright by Professor Jane Richardson. Reprinted with permission. [From G. G. Hammes, *Thermodynamics and Kinetics for the Biological Sciences*, Wiley-Interscience, New York, 2000.]

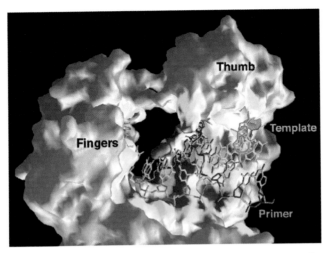

Figure 2-13. Structure of a catalytic fragment of DNA polymerase from *Bacillus stearothermophilus*. The "thumb" and "fingers" of the structure are labeled, and the template and primers can be seen as stick structures. Copyright by Professor L. S. Beese, Duke University. Reproduced with permission.

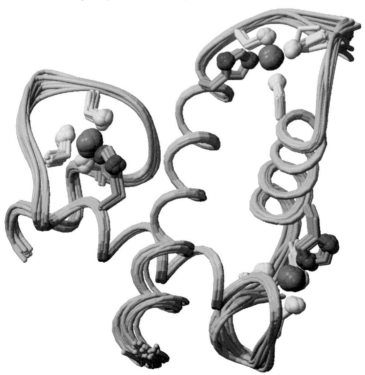

Figure 7-2. NMR structure of the TAZ2 (CH3) domain of CBP. A superposition of 20 structures are shown. The orange balls are Zn^{2+}, the yellow ball-and-stick representations are cysteines, and the purple and blue structures are imidazoles from histidines. The four α-helices can be easily discerned. PDB entry 1F81. Copyright by Professor David C. Richardson. Reprinted with permission. Kinemage graphics, then rendered in Raster3D.

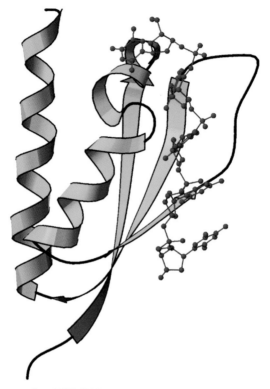

Figure 7-3. Structure of a KH3-DNA complex determined with NMR. The DNA (10mer) is shown in magenta as a ball-and-stick model, and the protein backbone is in cyan. PDB entry 1J5K. Copyright by Professor David C. Richardson. Reprinted with permission. Kinemage graphics, then rendered in Raster3D.

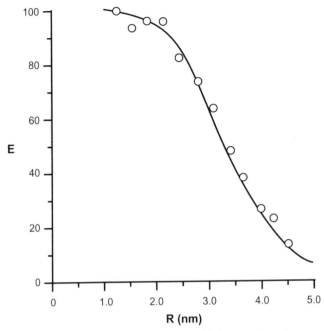

Figure 3-13. Per cent efficiency of energy transfer, E, in proline polymers of various lengths, R, with a fluorescence donor at one end of the polymer and a fluorescence acceptor at the other end. The solid line corresponds to Eq. 3-22. Reproduced with permission from L. Stryer and R. P. Haugland, *Proc. Natl. Acad. Sci. USA* **58**, 719 (1967).

shown in Figure 3-13, and the experimental values of R agreed well with those determined from molecular models. They also demonstrated that the dependence of R_0 on the degree of overlap of the donor emission with the acceptor absorbance was quantitatively correct.

With the establishment of the methodology in model systems, energy transfer measurements were made in many biological systems. The most critical part of the application is to label a protein or nucleic acid with a fluorescent probe at a specific site, with essentially no labeling at nonspecific sites. This is essential if a reliable interpretation of the data is to be made. As a specific example, consider the extensive studies on chloroplast coupling factor one, CF_1 (9). CF_1 is the soluble part of an enzyme that is critical for ATP synthesis in chloroplasts. Very similar enzymes are in all organisms. In animals, the enzyme is located in the mitochondria. The enzyme is responsible for the synthesis of ATP from ADP and inorganic phosphate, an important physiological reaction. The synthesis occurs when a pH gradient is established across the membrane, and protons are pumped across the membrane as the synthesis proceeds. Of course, an enzyme must catalyze the chemical reaction in both directions. Thus, the soluble portion of the enzyme catalyzes the reverse reaction, namely the hydrolysis of ATP. It cannot catalyze the synthetic reaction since a proton gradient cannot be created when the enzyme is not on the membrane.

CF$_1$ is a very complex protein and has five different subunits, with a subunit structure of $\alpha_3\beta_3\gamma\delta\varepsilon$. It binds three molecules of ATP or ADP, and the molecule is intrinsically asymmetric. In a series of experiments, fluorescent probes were put in a variety of positions on the molecule, and the distances between the probes were measured. Altogether, more than 30 distances were measured, including the distance of CF$_1$ sites from the surface of the membrane. These measurements were used to construct a model of the enzymes. The model is shown in Figure 3-14. At the time this work was done, the structure of the

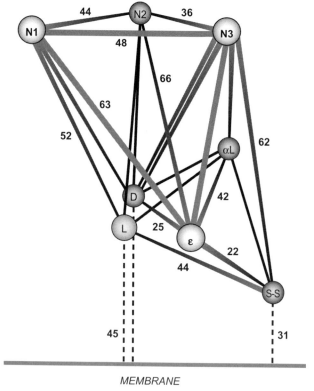

Figure 3-14. Spatial arrangement of specific sites on CF$_1$ determined by fluorescence energy transfer measurements. The sites shown are nucleotide binding sites (N$_1$, N$_2$, N$_3$), a disulfide on the γ polypeptide chain (S-S), specific sulfhydryl groups on the γ polypeptide chain (D, L), a sulfhydryl on the ε polypeptide chain (ε), and an amino group on an α polypeptide chain (αL). A few of the distances, in Å, are shown to establish the scale. The distances to the membrane are maximum values since the lines need not be perpendicular to the membrane surface. Reprinted from R. E. McCarty and G. G. Hammes, Molecular Architecture of Chloroplast Coupling Factor 1, *TIBS* **12**, 234 (1987). © 1987, with permission from Elsevier.

enzyme was unknown, and the model summarized some important features of the enzyme, including the distance between nucleotide binding sites and the intrinsically asymmetric nature of the structure. The asymmetric structure is critical for the coordination of the chemical reaction and the pumping of protons across the membrane. Some time later, the structure of the mitochondrial enzyme was determined by x-ray crystallography (10). The distances determined by energy transfer proved to be consistent with the crystal structure. In fact, the elucidation of structural features by energy transfer measurements is generally useful for complex structures for which the molecular structure has not yet been established.

DIHYDROFOLATE REDUCTASE

As an example of the use of fluorescence to study ligand binding to a protein and enzyme catalysis, we consider the enzyme dihydrofolate reductase (DHFR). DHFR catalyzes the reduction of 7,8-dihydrofolate by NADPH to give 5,6,7,8-tetrahydrofolate and $NADP^+$. Tetrahydrofolate is essential for the biosynthesis of purines, as well as several other important biological molecules. Because of its central position in metabolism, it has been extensively studied. It is also a target for anticancer drugs such as methotrexate. The structure of the enzyme and enzyme-substrate/inhibitor complexes are known and many mechanistic studies of the enzyme have been carried out (cf. references 11 and 12). This discussion will be limited to a few facets of the very extensive literature available for this enzyme.

NADPH is fluorescent, with an absorption maximum at about 340 nm and an emission maximum at about 450 nm. In addition, the enzyme contains tryptophan so that if the enzyme is excited at 290 nm, fluorescence is observed with a maximum intensity at 340–350 nm. If NADPH is added to the enzyme, this fluorescence is quenched about 65%, indicating the tryptophan environment is significantly altered when NADPH binds (13). Moreover, because NADPH has a maximum absorbance around 340 nm, significant energy transfer occurs from the excited tryptophan to bound NADPH. This results in a new emission maximum at 420–450 nm. The extent of NADPH binding to DHFR can be quantitatively assessed by measuring the quenching of tryptophan fluorescence. This can also be done by following the fluorescence at 450 nm. Quantitative analysis of the data permits the calculation of the equilibrium dissociation constant, which is 1.1 µM at pH 7.0.

The kinetics of ligand binding can also be measured by mixing NADPH and DHFR rapidly in a stopped flow apparatus and following the quenching of tryptophan fluorescence or the emission at 450 nm as a function of time. The time course of the reaction indicates a very rapid quenching, occurring in less than a second, followed by a relatively slower reaction, occurring in seconds. The first reaction is the binding of NADPH to DHFR, whereas the second reaction is the interconversion of the free enzyme from a form that does not bind NADPH to a form that does. The overall reaction can be represented by the following mechanism:

$$DHFR + NADPH \underset{k_{-1}}{\overset{k_1}{\rightleftharpoons}} DHFR-NADPH$$

$$k_{-2} \updownarrow k_2$$

$$DHFR'$$

A quantitative analysis of the data at pH 7.0 gives: $k_1 = 1.7 \times 10^6 M^{-1} s^{-1}$, $k_{-1} = 2.4 s^{-1}$. The equilibrium constant for the transition from DHFR to DHFR' was found to be about 1, with rate constants of approximately $2.5 \times 10^{-2} s^{-1}$ (k_2 and k_{-2}; 12). Note that for this mechanism, the overall equilibrium dissociation constant is

$$[(DHFR) + (DHFR')](NADPH)/(DHFR-NADPH)$$
$$= (k_{-1}/k_1)(1 + k_2/k_{-2}) = 2.8 \, \mu M$$

The dissociation constant determined from kinetics is only in fair agreement with the dissociation constant determined from direct binding studies.

The overall time course of the enzymatic reaction can be easily monitored by following the absorbance change at 340 nm because NADPH has an absorption peak at this wavelength whereas $NADP^+$ does not. In addition, the individual steps in the catalytic process can be characterized by stopped-flow studies that monitor alterations in the tryptophan fluorescence as the reaction proceeds. A detailed description of these experiments is beyond the scope of this discussion. However, it was possible to determine which substrate binds first to the enzyme, the rate constants for the binding steps, the rate constants for conformational changes within the enzyme-substrate complex, and the rate constants for hydride transfer. This permitted the development of a detailed mechanism for the reaction (13).

Another unique use of fluorescence is fluorescence microscopy. With this technique individual molecules are labeled with fluorescent tags and can be observed with a light microscope (cf. reference 14). Of course, it is not the molecule that is seen: it is the light from the molecule. This is analogous to seeing the light from a star, rather than the star itself. With this technique, kinetic and equilibrium properties of individual molecules can be studied. For example, the kinetics of binding of substrates and inhibitors to DHFR have been investigated by single-molecule fluorescence microscopy (15).

REFERENCES

1. K. Rosenheck and P. Doty, *Proc. Natl. Acad. Sci. USA* **47**, 1775 (1961).

2. G. G. Hammes, *Thermodynamics and Kinetics for the Biological Sciences*, Wiley-Interscience, New York (2000), pp. 59–68.

3. R. E. Dickerson and I. Geis, *Hemoglobin: Structure, Function, Evolution and Pathology*, Benjamin/Cummings Publ., Redwood City, CA (1983).

4. D. G. Anderson, G. G. Hammes, and F. G. Walz Jr., *Biochemistry* **7**, 1637 (1968).

5. L. J. Roman and S. C. Kowalczykowski, *Biochemistry* **28**, 2863 (1989).

6. L. J. Roman, A. K. Eggleston, and S. C. Kowalczykowski, *J. Biol. Chem.* **267**, 4207 (1992).

7. T. Förster, *Discussions Far. Soc.* **27**, 7 (1959).

8. L. Stryer and R. P. Haugland, *Proc. Natl. Acad. Sci. USA* **58**, 719 (1967).

9. R. E. McCarty and G. G. Hammes, *TIBS* **12**, 234 (1987).

10. J. Abrahams, A. Leslie, R. Lutter, and J. Walker, *Nature* **370**, 621 (1994).

11. M. R. Sawaya and J. Kraut, *Biochemistry* **36**, 586 (1997).

12. C. A. Fierke, K. A. Johnson, and S. J. Benkovic, *Biochemistry* **26**, 4085 (1987).

13. P. J. Cayley, S. M. J. Dunn, and R. W. King, *Biochemistry* **20**, 874 (1981).

14. S. Weiss, *Science* **283**, 1676 (1999).

15. Z. Zhang, P. T. R. Rajagopalan, T. Selzer, S. J. Benkovic, and G. G. Hammes, *Proc. Natl. Acad. USA* **101**, 13481 (2004).

PROBLEMS

3.1. Colored pH indicators are dyes that have different spectra for different ionization states. Assume the ionization of a pH indicator with pK = 6.00 can be written as:

$$HIn \rightleftarrows In^- + H^+$$

The measured absorbance in a 1-cm cell at an indicator concentration of 2×10^{-5} M is given in the table below.

λ (nm)	A at pH 3.00	A at pH 9.00
400	0.200	0.000
420	0.300	0.030
440	0.150	0.150
460	0.000	0.200
480	0.000	0.150

For the same concentration of indicator, an absorbance of 0.100 is measured at $\lambda = 400$ nm.

a. What is the pH of the solution?

b. What is the absorbance at $\lambda = 440$ nm?

c. At pH 7.00, for an indicator concentration of 3.00×10^{-5} M and a 1-cm cell, what is the absorbance at $\lambda = 400, 440,$ and 480 nm?

3.2. A sample of RNA is hydrolyzed and separated into three fractions by column chromatography. Two of the three fractions are pure nucleotides, but the third contains both adenylic and guanylic acid. At pH 7.0, the absorbance of the mixture is 0.305 at 280 nm and 0.655 at 250 nm in 1-cm cells. The molar extinction coefficients for each pure component at pH 7.0 are

	ε_{280} $(M^{-1}cm^{-1})$	ε_{250} $(M^{-1}cm^{-1})$
Adenylic acid	2,300	12,300
Guanylic acid	9,300	15,700

Calculate the mole ratio of adenine to guanine in the RNA.

3.3. The enzyme alcohol dehydrogenase catalyzes the oxidation of alcohol by NAD^+ to give acetaldehyde and NADH. NADH has an absorption maximum at 340 nm with an extinction coefficient of $6.20 \times 10^3 M^{-1} cm^{-1}$, whereas NAD^+ and the other reactants do not absorb significantly at 340 nm. Consequently, the oxidation reaction can be conveniently monitored by following the increase in absorbance at 340 nm.

a. What is the rate of production of NADH if it is observed that that the rate of absorbance increase in a 1-cm path length cell is 0.05/min?

b. When an excess of enzyme is added to NADH, the absorbance at 340 nm decreases by 13%. What is the extinction coefficient of NADH at 340 nm when it is bound to the enzyme?

c. The difference in extinction coefficient can be used to determine the binding constant for the association of NADH and the enzyme. The following data were obtained for the difference absorbance when NADH is added to 20 μM enzyme in a 1-cm path length cell.

$(NADH)_{total}$, μM	$\Delta A (340 \, nm)$
8.33	0.004
13.50	0.006
20.0	0.008
45.0	0.012

Calculate the binding constant from these data.

3.4. When complementary strands of deoxy-oligonucleotides are mixed, they form a double-stranded DNA at low temperatures but dissociate to single strands as the temperature is raised. This can be observed by a decrease in the absorbance of the solutions at 260 nm. The equilibrium for this process can be written as

$$D \rightleftarrows S_1 + S_2$$

with an equilibrium constant $K = (D)/[(S_1)(S_2)]$. Data are given below for two experiments. In one case, equal length strands and concentrations of oligoA and oligoT were mixed, whereas, in the other case, the same length strands and concentrations of oligoG and oligoC were mixed.

T(°C)	A (oligoA-T)	T(°C)	A(oligoG-C)
5	0.711	50	0.780
10	0.720	55	0.785
15	0.732	60	0.812
20	0.740	65	0.836
25	0.767	70	0.862
30	0.801	75	0.880
35	0.846	80	0.896
40	0.874	85	0.924
45	0.891	90	0.948
50	0.903	95	0.975
55	1.003		

a. Assume the molecules are duplexes at the lowest temperatures for which data are given and single strands at the highest temperatures. Calculate the fraction present as duplex at each temperature. Determine the melting temperature of each duplex, that is, the temperature at which half of the duplex has been converted to single strands.

b. What do these results tell you about the relative stability of A-T and G-C pairs?

c. Estimate the melting temperature of complementary oligonucleotides that are 50% A-T and 50% G-C, with the same length as in the above experiments. (Your result is not exact, but methods are available for calculating the melting temperatures of short DNA.)

3.5. Pyrenylmaleimide is a convenient fluorescent probe for labeling sulfhydryl residues on proteins. The fluorescence lifetime of pyrene on a protein sulfhydryl was measured by determining the relative fluorescence intensity after excitation of pyrene with a flash lamp. Typical data are given below.

Relative Fluorescence	Time (ns)
0.716	20
0.513	40
0.367	60
0.264	80
0.189	100

a. Determine the fluorescence lifetime.

b. If the quantum yield is 0.7, what is the natural lifetime.

c. When a ligand is bound to the protein, the fluorescence lifetime is 10% shorter. What might be the cause of this change?

3.6. Bovine rhodopsin is a photoreceptor protein that is an integral part of the disc membranes of retinal rod cells and plays a key role in vision. It has a tightly bound 11-*cis*-retinal that has a strong absorbance at about 500 nm. As part of the vision cycle, the retinal is bleached by conversion to the *trans* isomer. The protein has a molecular weight of 28,000–40,000. Three sites were labeled on the protein with fluorescent probes, sites A, B, and C. Fluorescence resonance energy transfer was measured between these three sites and the retinal and between the three sites themselves [C.-W. Wu and L. Stryer, *Proc. Natl. Acad. Sci. USA* **69**, 1104 (1972)]. Some of the results are summarized in the table below.

Engery Donor	Energy Acceptor	Transfer Efficiency	$R_0(\text{Å})$
A	11-*cis*-retinal	0.09	51
B	11-*cis*-retinal	0.36	52
C	11-*cis*-retinal	0.12	33
A	B	0.90	51
A	C	0.92	48
B	C	0.92	47

a. Calculate the distances between these six sites.

b. A protein of molecular weight 28,00–40,000 that is spherical has a radius of 40–45 Å. What can you say about the shape of rhodopsin? Sketch a model for the molecule based on the distances that have been measured.

CHAPTER 4

CIRCULAR DICHROISM, OPTICAL ROTARY DISPERSION, AND FLUORESCENCE POLARIZATION

INTRODUCTION

Most biological molecules possess molecular asymmetry, that is, their mirror images are not identical. Such molecules are said to be *chiral*. Probably the most well-known example is a carbon atom that is tetrahedrally bonded to four different atoms or groups of atoms. This example is described in organic chemistry textbooks and will not be dwelled on here. Less obvious examples of importance in biology are macromolecules that possess chirality. For example, helices can be wound in a left- or right-hand sense. The most common helix in proteins, the α-helix, is wound in a right-hand sense. Although most polynucleotides are wound in a right-hand sense, helices that wind in a left-hand sense also exist.

Chiral molecules can be distinguished by their interactions with polarized light. As we have discussed in Chapter 1, a light wave can be described as an electromagnetic sine wave with a characteristic frequency. In this description, a wave generating an electric field is perpendicular to a wave generating a magnetic field. If light is *unpolarized*, these waves are oriented randomly in space, as shown schematically in Figure 4-1. Looking along the electromagnetic wave, the random orientations look like the spokes of a wheel. For *linearly*, or *plane*, *polarized* light, the wave is oriented in only one direction, as depicted in Figure 3-1. Although the polarized light in the figure looks like a line (linearly polarized), the electric field wave is projected as a plane (plane

Spectroscopy for the Biological Sciences, by Gordon G. Hammes
Copyright © 2005 John Wiley & Sons, Inc.

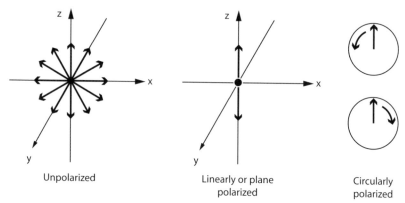

Figure 4-1. Diagrams of the electric field components of unpolarized, linearly or plane polarized, and circularly polarized light. The light is moving along the y axis. The arrows indicate the directions of the electric field. For unpolarized light all directions occur, whereas for linearly or plane polarized light only the z direction is found. For circularly polarized light, the direction of rotation can be clockwise or counterclockwise.

polarized) so that the two terms are used interchangeably. When a chiral molecule interacts with plane polarized light, the plane is rotated, with the direction and amount of rotation depending on the characteristics of the molecule.

Circularly polarized light sweeps out a circle as the electric wave propagates. The circle can be either right handed or left handed, as shown in Figure 4-1. The convention used is that if the circle moves clockwise, it is right handed, whereas if it moves counter-clockwise, it is left handed. Again, the interaction of circularly polarized light with chiral molecules will alter the circularly polarized light. Elliptically polarized light is also used and is essentially the same as circularly polarized light, but an ellipse is swept out by the electric field wave instead of a circle. We will not dwell on how polarized light is produced, except to note that light can be polarized when it is passed through certain materials. A pair of Polaroid sunglasses is a good example of how light can be polarized.

The interaction of a chiral molecule with polarized light is very specific and has proved to be an important method for characterizing both small molecule and macromolecular structures. Small molecule examples should be already well known to you. Amino acids in most biological systems are *levo* rotary (L), that is, they rotate plane-polarized light to the left, and sugars have various optical isomers. Enzymes, in fact, can distinguish between optical isomers and typically will only react with a single or a restricted group of small molecule isomers. Essentially two types of measurements are commonly made to determine the effects of molecules on polarized light, optical rotation and circular dichroism (CD). Optical rotation is a measure of the rotation of linearly polarized light by a molecule, and the wavelength dependence of the optical rotation is called optical rotary dispersion (ORD). CD, on the other hand, is the difference in absorption of left-hand and right-hand circularly polarized light. These effects are relatively small but can be measured readily with modern instrumentation. For a typical protein or nucleic acid with a 100 µM solution of chromophore, the polarized light plane with a wavelength around the electronic absorption maxima is rotated about 0.001–0.1 degrees for a sample 1 cm thick. For CD, the difference in absorption coefficients is about 0.03–0.3% of the total absorption.

OPTICAL ROTARY DISPERSION

The first studies of the optical properties of biological macromolecules were done with ORD because adequate equipment was not available for CD measurements. At the present time, the opposite is true: CD is the method of choice. Nevertheless, it is instructive to discuss both methods. As we shall see later, they are intimately related.

A typical experimental setup for measuring ORD is shown in Figure 4-2. The principle is that plane polarized light is passed through the sample, and the rotation of the plane of polarization of the light by the sample is measured. The analyzer for the emergent beam is an element that polarizes the beam. It

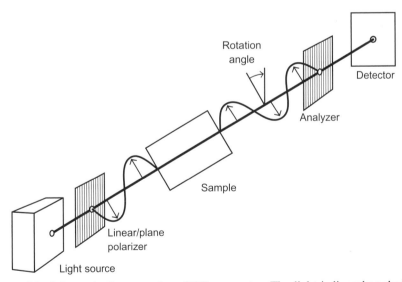

Figure 4-2. Schematic diagram of an ORD apparatus. The light is linearly polarized and the polarization plane is rotated by the sample. A second polarizer, the analyzer, can be rotated to obtain the maximum (or minimum) light intensity. The angle of rotation of the analyzer is a direct measure of the angle of rotation by the sample. For CD, circularly polarized light is used, and the difference in absorbance between left- and right-circularly polarized light is measured.

can be rotated until the light intensity after the analyzer disappears. Its polarization axis is then perpendicular to the direction of polarization of the beam emerging from the sample. The amount of rotation required for this disappearance is a direct measure of how much the original beam was rotated by the sample. Clockwise rotation is assigned as positive rotation and counterclockwise as negative rotation. The physical origin of this rotation is the fact that the sample is *circularly birefringent*, that is, the refractive index of the sample is different for left-hand circularly polarized light and right-hand circularly polarized light. A plane polarized beam of light can also be described as two opposite circularly polarized beams. A difference in the refractive index means that the speed at which the light passes through the sample is different for right-hand and left-hand circularly polarized light. This means the two polarized beams get out of phase as they pass through the sample. This phenomenon is described in Chapter 1. This phase difference is manifested as a rotation of the light.

If the measured angle of rotation is α, then the specific rotation, $[\alpha]$, is defined as

$$[\alpha] = \alpha/(dc) \tag{4-1}$$

where c is the concentration in g/cc and d is the path length in decimeters. The origin of this definition lies in the history of optical rotation: the first measurements of optical rotation were made in cells that were 10 cm (1 dm) long. Instead of the specific rotation, data are often reported as the molar rotation, $[\phi]$, which is defined as

$$[\phi] = 100\alpha/(lM) \tag{4-2}$$

where l is the path length in centimeters and M is the concentration in moles/liter. (For macromolecules, the concentration is often given in terms of the monomers.)

CIRCULAR DICHROISM

Materials display CD because the absorbance of left and right circularly polarized light is different for molecules with molecular asymmetry. One measure of the CD is simply the difference in absorbance of the material for left and right circularly polarized light, ΔA:

$$\Delta A = A_L - A_R \tag{4-3}$$

Circular dichroism can arise only in the spectral region where absorbance occurs—if the absorbance is essentially zero, there cannot be a measurable difference. For ORD, on the other hand, large rotations are measured in the regions where the sample absorbs light, but measurable rotation also occurs when the absorbance is essentially zero. This is helpful, for example, when looking at the optical activity of substances such as sugars, for which the absorbance is in the far ultraviolet. The large optical rotations in regions of high absorption are called Cotton effects.

As previously discussed, optical rotation arises because of the phase shift of the circularly polarized light that leads to circular birefringence. Circular dichroism arises because not only does the phase shift, but there is a differential decrease in the amplitude for right and left circularly polarized light. This leads to elliptical polarization. As with optical rotation, circular dichroism can be either positive or negative. Circular dichroism is reported either as a differential extinction coefficient,

$$\Delta\varepsilon = \varepsilon_L - \varepsilon_R = \Delta A / l M \tag{4-4}$$

or more often as the ellipticity, θ:

$$\theta = 2.303 \Delta A \, 180 / (4\pi) \text{ degrees} \tag{4-5}$$

As before, the molar ellipticity is

$$[\theta] = 100 \theta / l M \tag{4-6}$$

Despite this definition of $[\theta]$, it is usually reported in the units of deg cm^2 dmol^{-1} because of the historical precedents. The differential extinction coefficient and the specific rotation are related by the relationship

$$[\theta] = 3300 \Delta\varepsilon \tag{4-7}$$

The numerical coefficient follows from the Beer-Lambert law for absorption and the various relationships in previous equations.

Since ORD and CD arise from the same physical phenomena, namely the effect of molecular asymmetry on polarized light, one might imagine that the two should be closely related. In fact, these two measurements are not independent: if the optical rotation is known, the circular dichroism can be calculated and vice versa. Although the formal equations for these transformations are well known, in practice the mathematics are sufficiently difficult so that this transformation is rarely done. In fact, CD is the experimental method of choice for a variety of reasons, mostly because circular dichroism is only found at absorption bands so that the interpretation of the spectra is somewhat easier.

OPTICAL ROTARY DISPERSION AND CIRCULAR DICHROISM OF PROTEINS

The simplest model for the optical activity of a protein is to assume that the optical activity is the sum of the optical activity of the individual amino acids.

This model is clearly incorrect: the sum of the optical activity of the amino acids is usually very small relative to the measured optical activity of a protein. This observation is not surprising: we saw previously that the ultraviolet spectra of proteins were primarily due to the peptide bonds. Similarly, the optical activity of proteins is primarily due to the macromolecular structure itself. In fact, specific protein structures have characteristic ORD and CD spectra.

Three of the fundamental structures of proteins are the α-helix, the β-sheet, and the random coil (cf. reference 1 and Chapter 2). The α-helix is a right-hand helix stabilized by short-range hydrogen bonds between backbone peptide bonds, whereas the β-sheet structure is composed of parallel polypeptide chains, also stabilized by hydrogen bonds between the backbone peptide bonds (Figs. 2-8 and 2-9, see color plates). The random coil is envisaged as a random arrangement of the backbone although it is rarely random—irregular might be a better term. The ORD and CD spectra of many different homo-polypeptide chains have been determined under experimental conditions where the structures are known. The results are summarized in Figure 4-3 (2). Within relatively small deviations, the spectra are the same for the common structures of the homo-polypeptides. Note that, as expected, the CD is centered around the ultraviolet absorption spectra of the proteins. Also note that the spectra are reasonably distinct for these three different structures.

If the assumption is made that only α-helix, β-sheet, and random coil structures are present and that these three structures have known optical proper-

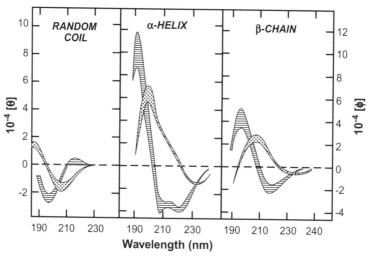

Figure 4-3. ORD (*dotted*) and CD (*crosshatched*) of homopolypeptides in the random coil, α-helix, and β-sheet conformations. The range of values found is indicated by the shaded areas. Reproduced with permission from W. B. Gratzer and D. A. Cowburn, *Nature* **222**, 426 (1969).

ties, it should be possible to develop algorithms to deconvolute the ORD and CD spectra of a given protein. The implicit assumption in these algorithms is that ORD and CD spectra of the three different structures are additive, that is, the presence of one structure does not influence the contributions from another structure. Thus measurement of the ORD or CD spectrum over the wavelength region of 175–250 nm would permit the calculation of the fraction of the protein structure present in each of the three conformations. A number of attempts have been made to do this, and the results have been compared with known crystal structures (cf. 3). In some cases, crystal structures have been used to derive a set of self-consistent parameters, assuming the crystal and solution structures are identical. Reasonable agreement between the calculated structures and the actual structures is often found, but the calculations are not quantitative for several reasons. First, the optical properties of homopolymers are not accurate predictors of the optical properties of the protein. The ORD and CD spectra depend on the microenvironment of each residue, so that they will certainly be dependent on the specific sequence and size of the protein. Second, protein structures are not so simple that their domains can be described by the three relatively simple structures assumed for the algorithms. Other structures that can influence CD and ORD spectra include various helices other than the α-helix, disulfide bonds, and various β-turns. Many of the fitting algorithms include structures other than the three discussed here. Finally, the side chains of amino acids influence the spectra—this is particularly true at wavelengths above about 250 nm where tyrosine, phenylalanine, and tryptophan have strong absorption bands. Prosthetic groups, such as hemes, would have a similar effect at the wavelength where they have strong absorption bands. Nevertheless, ORD and CD spectra are useful semiquantitative predictors of protein structure and also serve as good indicators of structural changes that may occur as experimental conditions are changed (cf. 4).

OPTICAL ROTATION AND CIRCULAR DICHROISM OF NUCLEIC ACIDS

The bases of nucleic acids are not intrinsically optically active; however, the sugars are optically active and can induce optical activity in the bases. This optical activity is relatively small, but as might be anticipated from the discussion of proteins, the ORD and CD spectra of nucleic acids are not just the sum of those of the monomers. In fact, CD and ORD are very good indicators of secondary structure such as helices and are the method of choice for following changes in secondary structure. As an example, consider polyadenylic acid (poly-A) (5). As shown in Figure 4-4, at neutral pH and room temperature, native poly-A has a very large ellipticity due to the fact that it exists in a helical structure. If the structure is destroyed by raising the temperature or going to extremes of pH, the ellipticity essentially disappears. Also shown in the figure is the ellipticity of adenylic acid, which is quite small. The

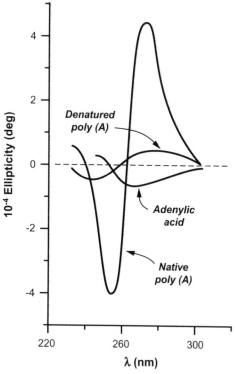

Figure 4-4. Circular dichroism of adenylic acid, denatured polyadenylic acid, and native polyadenylic acid. D. Freifelder, *Physical Biochemistry*, 2nd edition, W. H. Freeman, New York, 1982, p. 594. © 1976, 1982 by W. H. Freeman and Company. Used with permission.

ORD and CD of single-stranded polynucleotides have been extensively studied. The large ellipticity is primarily due to the interactions of nearest neighbors, that is, adjacent bases. They are stacked on top of each other, and this stacking is primarily responsible for the altered optical properties of the helical structures. To a good approximation, the ORD and CD properties of a base are influenced only by their nearest neighbors. Consequently, the spectra of polymers can be readily simulated by adding together the spectrum of each base, taking into account nearest neighbor interactions only.

The situation is much more complex for double-stranded nucleic acids such as DNA, and structural predictions are much more difficult, but CD and ORD are still useful tools. As an example, the CD of *E. coli* DNA is shown in Figure 4-5 for both the native and denatured forms (6). Although structural predictions from ORD and CD spectra are difficult, it is true, nevertheless, that specific structures of DNA and RNA have very characteristic spectra, and changes in the spectra are good monitors of structural changes (cf. 4).

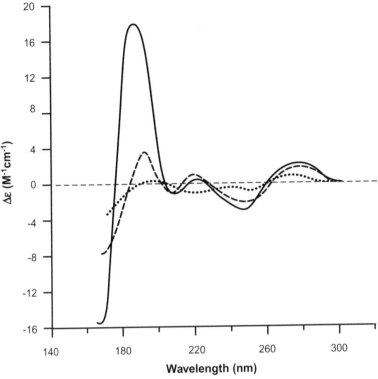

Figure 4-5. The CD of *E. coli* DNA in its native form at 20°C (—), heat denatured form at 60°C (--), and the average CD of the four deoxynucleotides (...). From C. A. Sprecher and W. C. Johnson, Jr., *Biopolymers* **16**, 2243 (1977). Reprinted with permission of John Wiley & Sons, Inc. © 1977.

SMALL MOLECULE BINDING TO DNA

DNA is an obvious target for drug intervention since disruption of its structure clearly will have significant biological implications. The negatively charged phosphate groups along the backbone structure of DNA can bind small molecules through electrostatic interactions, but the major sites of drug binding in double-stranded DNA are the major and minor grooves in the double helical structure and intercalation between the bases. These three modes of binding are shown schematically in Figure 4-6 (7). Many of the drugs used contain aromatic rings, and as a specific example of such binding we consider the interaction of DNA with acridine orange, even though acridine orange is not actually a drug (8).

The structure of acridine orange is shown in Figure 4-7. It is not optically active by itself but binding to DNA induces optical activity into the molecule.

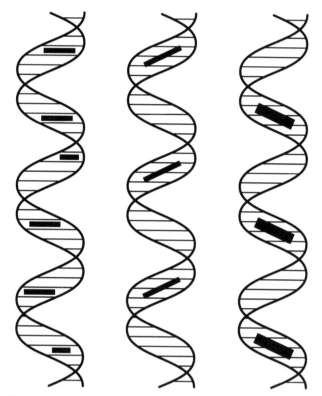

Figure 4-6. Schematic diagram of the intercalation of ligands between the bases, and in the minor and major grooves of a DNA double helix. From M. Ardhammar, B. Norden, and T. Kurucsev in *Circular Dichroism: Principles and Applications* (N. Berova, K. Nakanishi, and R. W. Woody, eds.), 2nd edition, John Wiley and Sons., New York, 2000, p. 746. Reprinted with permission of John Wiley & Sons, Inc. © 2000.

This is because binding causes the electronic energy states of achiral acridine orange to be coupled with the electronic energy states of chiral DNA. Induced optical activity is quite common when achiral small molecules bind to chiral macromolecules. The CD spectra of acridine orange bound to DNA are shown in Figure 4-7 for various ratios of (bound dye)/DNA (8). As expected, the optical activity is centered around the electronic absorption band of acridine orange, which has a maximum at about 430 nm. At very low values of (bound dye)/(acridine orange), a negative value of $\Delta\varepsilon$ is observed, but as this ratio increases, a large positive band develops. At the highest ratio shown, both strong negative and positive bands are observed. A molecular interpretation of these results has been developed. At very low concentrations of acridine orange, the binding is through intercalation between the bases, with the acridine orange ring structure parallel to the bases. This produces a negative

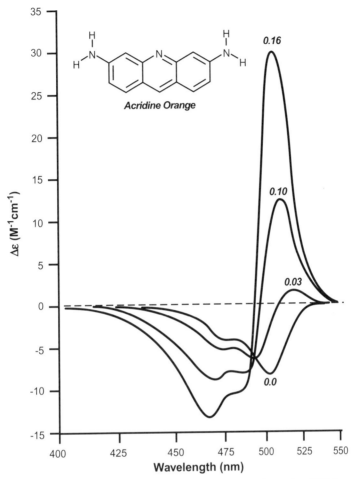

Figure 4-7. Circular dichroism per mole of acridine orange bound to DNA at the indicated values of [bound dye]/[DNA]. The total dye concentration is 10^{-5} M. Reprinted in part with permission from D. Fornasiero and T. Kurncsev, *J. Phys. Chem.* **85**, 613 (1981). © 1981 by American Chemical Society.

induced circular dichroism with a maximum $\Delta\varepsilon$ of about $-8 \, M^{-1} \, cm^{-1}$. As the concentration of the ligand increases, the dye binds to the groove which induces a positive CD with a maximum value of about $60 \, M^{-1} \, cm^{-1}$. These two induced CD signals are enhanced at higher dye concentrations due to interactions between the electronic energy levels of the bound dye molecules.

This example illustrates the exquisite sensitivity of the CD spectrum to the nature of the binding process and to the secondary structure of the DNA. Studies with many different types of ligands have been carried out (cf. 7).

PROTEIN FOLDING

Understanding protein structure in molecular terms is a long-standing goal of biochemistry. While great progress has been made toward this goal, we still cannot predict the three-dimensional structure of a protein from a knowledge of the one-dimensional sequence of amino acids that make up a protein. One approach to this end is to study the folding and unfolding of proteins with the rationale that understanding these processes will provide a better understanding of protein structure. In addition, understanding the folding of proteins is of great physiological significance since proteins are synthesized unfolded *in vivo* and must be folded into specific structures in order to perform their biological activities. Many reviews of protein folding are available (cf. 9), and only one particular example is presented here. CD and ORD are particularly useful tools for the study of protein folding and unfolding since they are extremely sensitive to secondary and tertiary structure of protein so that large changes are generally seen as a result of the folding process.

Traditionally, proteins were thought to be synthesized and then immediately folded into their biologically active form. However, in recent years a number of examples have emerged in which the unfolded form exists *in vivo* until it is folded into its biologically active form by specific conditions in the external environment such as the presence of a specific ligand. The biological significance is not understood, but the evidence for such structures is quite convincing. An example of this phenomenon is the protein associated with the ribozyme ribonuclease P (10). Ribonuclease P is a ribonucleic acid (RNA) catalyst. Its activity and/or specificity is enhanced by the binding of the catalytic RNA to a protein, and the protein is essential for *in vivo* catalysis (11). We will not discuss the eatalytic reaction itself which is the processing of the 5'-leader sequences from precursor tRNA. This ribozyme is quite ubiquitous in nature: in bacteria, the complete catalytic unit consists of a single RNA of about 400 nucleotides and a single protein of about 120 amino acids.

When the protein from *B. subtilis* is isolated, it is unfolded at physiological temperatures in the absence of ligands (12). This can be ascertained from the CD spectrum in sodium cacodylate (pH 7), which does not bind to the protein. The spectrum of the protein in the far and near ultraviolet is shown in Figure 4-8. The far ultraviolet spectrum is due to the protein background whereas the near ultraviolet spectrum is primarily due to tyrosine residues in the protein. Included in Figure 4-8 are the spectra in the presence of 10 or 20 mM sulfate. An obvious change in the spectrum occurs. The far ultraviolet spectrum is now consistent with the known crystal structure of the protein which contains α-helices and β-sheets, and the near ultraviolet spectrum changes because the electronic environments of the tyrosines are altered.

The change in structure can be quantified by measuring the change in the CD at 222 nm as a function of the ligand concentration. This is shown for a variety of ligands in Figure 4-9. A variety of anions cause the protein to go

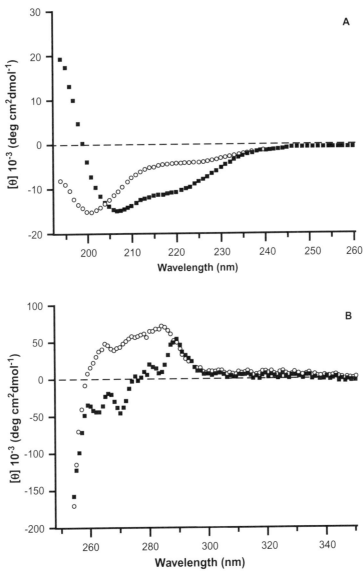

Figure 4-8. Circular dichroism of P protein in the presence (○) and absence (■) of 10 (B) or 20 (A) mM sulfate at pH 7.0 in 10 mM sodium cacodylate. Reprinted in part with permission from C. J. Henkels, J. C. Kurz, C. A. Fierke, and T. G. Oas, *Biochemistry* **40**, 2777 (2001). © 2001 by American Chemical Society.

from a disordered structure to a more folded structure, with highly charged anions such as polyphosphates being more effective than monovalent ions such as chloride. The stoichiometry of the ligand binding was determined for pyrophosphate and was found to be 2 pyrophosphates/protein. A simple model

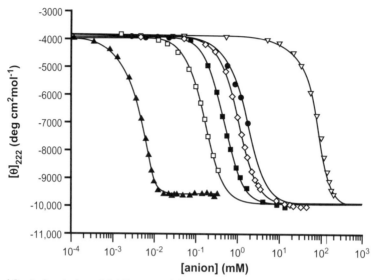

Figure 4.9. Anion-induced folding transitions of the P protein followed by the change in the CD signal at 222 nm at 37°C. The anions are dCTP (▲), CMP (□), sulfate (■), phosphate (◇), dCMP (•), and formate (▽). Selected data from C. J. Henkels, J. C. Kurz, C. A. Fierke, and T. G. Oas, *Biochemistry* **40**, 2777 (2001). Figure courtesy of Dr. C. J. Henkels, Duke University. Reproduced with permission.

that explains the observations is that the ligand binds to the folded protein but not the unfolded protein. The apparent equilibrium constant, K_{app}, for the ratio of the denatured (folded) protein, D, to the native state (folded), N, can be obtained directly from the data in Figure 4-9 if the assumption is made that only two forms of the protein exist, folded and unfolded. For example, the apparent equilibrium constant is unity when the change in the CD is half of the total change. The apparent equilibrium constant at any point on the curve is given by

$$K_{app} = (\theta_L - \theta_D)/(\theta_N - \theta_L) = \Sigma N/\Sigma D \quad (4\text{-}7)$$

where the θ's are the absolute values of the circular dichroism in the figure for a given ligand concentration, L, the denatured form, D, and the native form, N. The summation is over all liganded forms of the denatured and native proteins. If the ligand is assumed to bind to two independent and equivalent sites, the folding mechanism can be written as

$$D \overset{K_{fold}}{\rightleftarrows} N + 2L \overset{K_a}{\rightleftarrows} NL + L \overset{K_a}{\rightleftarrows} NL_2 \quad (4\text{-}8)$$

This mechanism gives

$$K_{app} = \frac{(N)+(NL)+(NL_2)}{(D)} = \frac{(N)[1+(NL)/(N)+(NL_2)/(N)]}{(D)}$$

$$= (N)/(D)\left[1+2K_a(L)+K_a^2(L)^2\right] = K_{fold}\left[1+K_a(L)\right]^2 \qquad (4\text{-}9)$$

where $K_{fold} = (N)/(D)$ and K_a is the is the microscopic association constant for the binding of ligand to the native protein [K_a = $(NL)/[2(N)(L)]$ = $2(NL_2)/[(NL)(L)]$, (1)]. The two constants K_{fold} and K_a cannot be determined independently from the data in Figure 4-9. However, an independent method, which will not be described here, gives an estimated value of 0.0071 for K_{fold} at 37°C. With this value in hand, it is possible to calculate the microscopic association constants for the ligands.

This example illustrates two important concepts. First, it shows that measurement of the CD provides a means of following the transition between the denatured and native state of proteins. Second, it shows how the binding of a ligand to the native state can convert the denatured protein to its native structure. (Coupled equilibria are common in biological systems.) These concepts are of both practical and theoretical value for studying and understanding biochemical processes.

INTERACTION OF DNA WITH ZINC FINGER PROTEINS

Zinc finger motifs are found in proteins that regulate transcription. As the name implies, the structure is a finger consisting of a loop of β-structure with two cysteines and an α-helix with two histidines. The four side chains of these amino acids tightly bind a zinc ion, as shown schematically in Figure 4-10. Transcription factors contain multiple "fingers," and the α-helix of each finger binds in the major groove of double-stranded DNA, thereby regulating transcription. The interaction of zinc finger proteins with DNA has been extensively studied. For example, the interaction of a transcription factor that regulates the expression of the protein metallothionein in mouse and human cells with DNA has been studied using CD (13, 14). The transcription factor contains six zinc fingers, which were isolated from the rest of the transcription factor. Preparations contained three to six-fingers. The six-finger protein binds three zinc ions very tightly, and far ultraviolet CD spectra showed that the conformation of the protein was unchanged once the three tight binding sites were occupied by zinc, although three additional zinc ions were bound more weakly.

The binding of the isolated zinc fingers to a 23-mer DNA that contains the sequence involved in the regulatory process was studied with CD. As the number of zinc fingers was increased from 3 to 6, a negative ellipticity at about 210 nm decreased in magnitude (approached zero) with a concomitant increase in a positive ellipticity at 280 nm. This change is not due to the protein, whose CD spectrum was subtracted from the experimentally determined spectrum. It is due to a change in the DNA structure from the B to A form. The

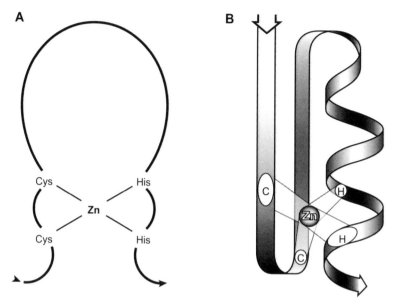

Figure 4-10. Schematic drawing of the Zn-finger structure. A. Coordination of Zn to cysteine and histidine ligands, B. Secondary structure. Reprinted from R. M. Evans and S. M. Hollenberg, Zinc Fingers: Gilt by Association, *Cell* **52**, 1(1988). © 1988, with permission of Elsevier.

B form of DNA is the structure stable at high humidity and is the form most prevalent in cells, whereas the A form is stable at lower humidity. The A form is characterized by a tilting of the base pairs relative to the axis of the helix and a change in the position of the sugars. This change in DNA structure appears to be a general consequence of binding zinc fingers and is thought to be involved in the regulation of transcription. The A and B forms of DNA have somewhat different CD spectra, which make it possible to interpret the CD changes associated with zinc finger binding in structural terms.

The interaction of many different proteins and nucleic interactions have been probed with CD (13). These interactions are of great physiological importance, and CD has proved to be a valuable tool for probing the structural changes that occur.

FLUORESCENCE POLARIZATION

With the introduction of the concept of polarized light, we can introduce an interesting phenomenon observed with fluorescence. Normal fluorescence experiments are not carried out with polarized light so the fluorescence observed also is not polarized. However, if the light used for excitation of fluorescence is polarized, then the fluorescence emitted is also often polarized. The basic idea is not hard to understand: if the electric field wave is oriented in space, it will interact with the electric dipole of the molecule, and this interaction can ultimately lead to the absorption of energy and fluorescence. The polarization of the light emitted will depend on the relative orientation of the electric field and the electric dipole and how much it changes during the fluorescence process. (The polarization of the emitted light is determined by the "transition dipole" which is generally oriented differently than the electric dipole.) Now let us consider what happens with a population of molecules interacting with polarized light.

The experiment itself is straightforward. A polarizer is put in front of the excitation beam so that molecules are excited with polarized light. A second polarizer is put in front of the detector for the emitted light. Two measurements are made: the intensity of the emitted light with the axis of the detection polarizer parallel to the polarization axis of the excitation polarizer, $I_{||}$, and the intensity when the two axes are perpendicular, $I_{\perp}$. The polarization, P, is defined as:

$$P = (I_{||} - I_{\perp})/(I_{||} + I_{\perp}) \tag{4-10}$$

If the emitted light is completely unpolarized, the parallel and perpendicular intensities would be identical and the polarization would be zero. This would happen if the molecules rotate very rapidly during the time that the fluorescence occurs. Since typical fluorescence lifetimes are in the range of 1–100 ns, this would mean that the molecules would have to rotate many times during this relatively short lifetime. This is, in fact, what happens for small molecules where polarization is usually not observed. The opposite extreme is the case when the molecules do not rotate at all during the lifetime of the fluorescence but are randomly oriented in space. This would be the situation, for example, in extremely viscous solvents. The calculation for this situation is somewhat complex, but can be done. For this limit, the polarization is $1/2$. For some macromolecules and ligands bound to macromolecules, the rate of rotation of the fluorescent species is comparable to the fluorescent lifetime. For such cases, the polarization lies between 0 and $1/2$, and measurement of the polarization gives information about the rate of rotation of the fluorophore.

The time dependence of the polarization will provide quantitative information about the rate of rotation of fluorescent molecules. Usually the data are analyzed in terms of the anisotropy, A, rather than the polarization, although the two are conceptually equivalent. It turns out the anisotropy is more easily related to rotational motion.

$$A = (I_{||} - I_{\perp})/(I_{||} + 2I_{\perp}) \tag{4-11}$$

If the time dependence of the fluorescence is measured for the parallel and perpendicular component, the time dependence of the polarization can be calculated. These are not easy measurements since fluorescence decays in nanoseconds. Theoretical considerations show that for a simple rotation

$$A(t) = A(0)\exp(-t/\tau_c) \tag{4-12}$$

where t is the time, $A(0)$ is a constant, and τ_c is the rotational correlation time and is related to the rotational diffusion constant, D_{rot}, by $\tau_c = 1/(6D_{rot})$. For example, the rotational correlation time for a fluorescent derivative of chymotrypsin is 15 ns, which is typical for small proteins. The rotational correlation time becomes longer as the molecular volume of the macromolecule increases, so that it is a direct measure of the size of the macromolecule. The exact relationship is complex, as both molecular weight and shape are important. If more than one mode of rotation is possible, multiple exponential decays may be observed.

Polarization and/or anisotropy provide information about the orientation and rotational freedom of fluorescent molecules. This, in turn, can often provide useful information about biological structures and mechanisms.

INTEGRATION OF HIV GENOME INTO HOST GENOME

HIV is a retrovirus that is unfortunately well known to everyone. In order for the virus to infect its host, the HIV genome must be integrated into the host genome. This integration is quite complex: it involves processing the HIV DNA, strand transfer, and DNA repair. Integrase is the enzyme responsible for the 3'-processing and strand transfer. These reactions can be monitored *in vitro* with short oligonucleotides that are models for the end of the viral DNA. The catalytic core of the enzyme recognizes a specific sequence of DNA. The self-association of integrase appears to be important for its function. The oligomeric state of integrase bound to viral DNA can be monitored through time-resolved fluorescence anisotropy measurements (15). It is a useful tool for this purpose because very low protein and DNA concentrations can be monitored, and the rotational correlation time is directly related to the size of the macromolecule.

The rotational correlation time for integrase was determined by monitoring the time decay of tryptophan fluorescence. The rotational correlation time decreased from about 90 ns to 20 ns at 25°C as viral-specific DNA, a 21 base sequence, was added. This was attributed to association of integrase with the DNA. The concentration of integrase was only 100 nM in these experiments. Independent studies of integrase established that the rotational correlation time in the absence of DNA was associated with a tetrameric structure of the protein. The actual situation is a bit more complex than this, as some other aggregates are present, but the dominant species is a tetramer, which becomes even more predominant at 37°C. The rotational correlation time of the 21-mer DNA duplex was determined by modifying the DNA with fluorescein. The dominant rotational correlation time is about 2 ns. A much faster rotational correlation time is also observed due to rotation of fluorescein within the DNA. Multiple rotational correlation times are commonly observed so that is necessary to establish what is being measured on a molecular basis. As the fluorescein-labeled DNA was titrated with integrase, an additional rotational correlation time was observed of about 15 ns. This rotational correlation time is attributed to the complex formed, and is consistent with that determined from measurement of the correlation time associated with the tryptophan fluorescence. The rotational correlation time for monomer integrase, molecular weight 32,000, is 16 ns.

These elegant experiments with the integrase-DNA complex, and the many control experiments, demonstrate that the integrase depolymerizes when it binds to DNA. Furthermore, the predominant binding state at 25°C is monomeric, whereas at 37°C a mixture of monomers and dimers is present. This raises the interesting question as to what is the active oligomeric state of integrase? Prior to this study, the prevailing mechanism was thought to involve a multimeric structure of integrase. These results suggest that *in vitro*, monomers and dimers may be enzymatically active, although the *in vivo* activity, which is considerably more complex, probably involves structures larger than dimers.

α-KETOGLUTARATE DEHYDROGENASE

The α-ketogluarate dehydrogenase complex from *E. coli* contains three enzymes that catalyze the overall reaction

$$\alpha\text{-ketogluarate} + CoA + NAD^+ \rightarrow \text{succinyl-CoA}$$
$$+ CO_2 + NADH + H^+ \tag{4-13}$$

The first enzyme decarboxylates the α-ketoglutarate. The intermediate formed with thiamine pyrophosphate transfers the succinyl moiety to lipoamide, and this intermediate is oxidized to form a succinyl-lipoic acid intermediate. The succinyl group is transferred to CoA, and finally the dihydrolipoamide is oxidized. This last reaction involves an enzyme-bound flavin and reduces NAD^+ to NADH. The multienzyme complex has a molecular weight of about 2.5×10^6 and contains 12 copies of the first enzyme, 24 copies of the second, and 12 copies of the third (16). The intermediates in the reaction sequence are bound to lipoamide, that is, lipoic acid covalently attached to a lysine through an amide linkage. The structure of lipoic acid is shown in Figure 4-11. It can exist in an oxidized form with a disulfide at the end of the chain, or in a reduced form with two sulfhydryl groups. The proposed mechanism involves the reduced lipoic acid-bound intermediates moving between the active sites of the three enzymes. Two questions arise: are the active sites close enough to

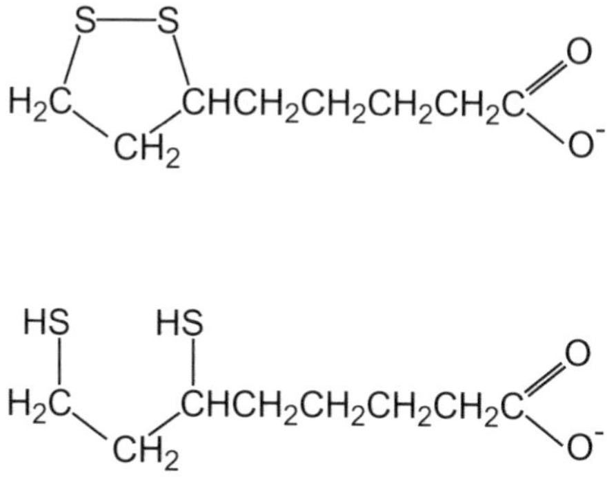

Figure 4-11. Structures of oxidized and reduced lipoic acid.

permit lipoic acid to span the distances between the active sites, and does the lipoic acid rotate between the active sites fast enough for the reaction to proceed efficiently? Answers to both of these questions can be obtained with fluorescence methodology.

Fluorescence resonance energy transfer measurements between the active sites of the three enzymes show they are about 3 nm apart, approximately the maximum span of a lipoic acid (17). Other experiments showed that succinyl and electron transfer between lipoic acids can occur, so that the potential distance an intermediate can be transferred is longer than a single lipoic acid. Thus the energy transfer measurements demonstrated that the transfer of intermediates occurs over a relatively long distance and that lipoic acid is a viable intermediate.

Does the lipoic acid rotate fast enough to serve the postulated transfer function? This is not an easy question to answer because no easy method exists for measuring the rate of rotation. However, fluorescence anisotropy measurements provide some insight into this rate (18). The lipoic acid was labeled with a fluorescent probe on the sulfhydryl that normally carries a reaction intermediate. The fluorescence lifetime of the probe and the dynamic fluorescence anisotropy were then measured. A typical time course for the fluorescence and anisotropy is shown in Figure 4-12. As might be expected, the dynamic fluorescence anisotropy is quite complex. Three components are found. One has a very long rotational correlation time and is due to the overall rotation of the very large multienzyme complex. A second component has a rotational correlation time of about 25 ns and is due to local rotation of the fluorescent probe in a hydrophobic environment. Rotational correlation times of this magnitude have been observed often for local motion of ligands bound to macromolecules. The third component has a rotational correlation time of about 350 ns. This can be attributed to rotation of lipoic acid between the catalytic sites. This component of the anisotropy decay is not seen if the enzyme to which the lipoic acid is bound is separated from the other two enzymes. It cannot be

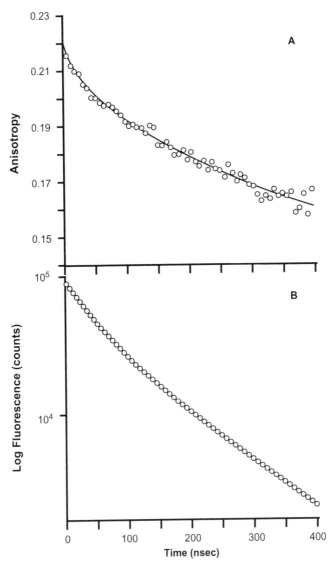

Figure 4-12. Time course of the fluorescence (B) and anisotropy (A) decay of pyrene covalently attached to the lipoic acid of *E coli* α-ketoglutarate dehydrogenase. Reprinted in part with permission from D. E. Waskeiwicz and G. G. Hammes, *Biochemistry* **22**, 6489 (1982). © 1982 by American Chemical Society.

ascertained whether the rate-limiting process is the actual rotation or dissociation of the probe from one of the catalytic sites, but in any event these results establish that the rate of rotation is sufficiently fast to support the observed catalytic rate, namely a turnover number of about $130\,s^{-1}$.

The uses of fluorescence anisotropy to probe biological mechanisms discussed are intended as illustrative examples of the unique information that can be obtained. They also illustrate the difficulty of interpreting the results that are obtained. As with any physical method, great care must be taken to explore all possible explanations and careful control experiments are required.

REFERENCES

1. G. G. Hammes, *Thermodynamics and Kinetics for the Biological Sciences*, Wiley-Interscience, New York, 2000.
2. W. B. Gratzer and D. A. Cowburn, *Nature* **222**, 426 (1969).
3. N. Sreeramama and R. W. Woody, *J. Mol. Biol.* **242**, 497 (1994).
4. N. Berova, K. Nakanishi and R. W. Woody, eds, *Circular Dichroism: Principles and Applications*, 2nd edition, John Wiley & Sons, New York, 2000.
5. D. Freifelder, *Physical Biochemistry*, 2nd edition, W. H. Freeman, New York, 1982, p. 594.
6. C. A. Sprecher and W. C. Johnson Jr., *Biopolymers* **16**, 2243 (1997).
7. M. Ardhammar, B. Norden, and T. Kurucsev in reference 4, pp. 741–768.
8. D. Fornasiero and T. Kurucsev, *J. Phys. Chem.* **85**, 613 (1981).
9. J. K. Myers and T. G. Oas, *Annu. Rev. Biochem.* **71**, 783 (2002).
10. S. Altman and L. Kirsebom, *Ribonuclease P in the RNA World* (R. F. Gesteland, T. R. Cech, and J. F. Atkins, ed.), Cold Spring Harbor Laboratory Press, Plainview, NY, 1999, pp. 351–380.
11. R. Kole, M. F. Baer, B. C. Stark, and S. Altman, *Cell* **19**, 881 (1980).
12. C. H. Henkels, J. C. Kurz, C. A. Fierke, and T. G. Oas, *Biochemistry* **40**, 2777 (2001).
13. D. M. Gray in reference 4, pp. 769–796.
14. X. Chen, A. Agarwal, and D. P. Giedroc, *Biochemistry* **37**, 11152 (1998).
15. E. Duprez, P. Tanc, H. Leh, J.-F. Mouscadet, C. Auclair, M. E. Hawkins, and J.-C. Brochon, *Proc. Natl Acad. Sci. USA* **98**, 10090 (2001).
16. L. J. Reed, *Acc. Chem. Res.* **7**, 40 (1974).
17. K. J. Angelides and G. G. Hammes, *Biochemistry* **18**, 5531 (1979).
18. D. E. Waskiewicz and G. G. Hammes, *Biochemistry* **22**, 6489 (1982).

PROBLEMS

4.1. The sugar D-mannose can exist as two enantiomers that rotate light in opposite directions. The (+) configuration has a specific rotation of $29.3°$ d^{-1}g/cc at 589.3 nm, 20°C, whereas the (−) configuration has a specific rotation of $-17.0° d^{-1}$g/cc at the same wavelength. When either pure enantiomer is put into water, the optical rotation changes until a specific rotation of $14.2° d^{-1}$g/cc is reached. Calculate the ratio of (+) and (−) enantiomers at equilibrium.

4.2. Solutions of RNA and DNA at a concentration of 2.00×10^{-5} M nucleotides (monomers) have the following differential absorption characteristics in a 1-cm cell:

λ(nm)	$A_L - A_R$, DNA	$A_L - A_R$, RNA
240	-4.40×10^{-4}	0.00
260	0.00	12.0×10^{-4}
280	6.00×10^{-4}	3.20×10^{-4}
300	0.20×10^{-4}	-1.00×10^{-4}

  a. Calculate $\varepsilon_L - \varepsilon_R$ for the DNA and RNA at each wavelength.
  b. Calculate the molar ellipticity for DNA and RNA at each wavelength.
  c. A mixture of the DNA and RNA has the following differential absorption characteristics.

λ(nm)	$A_L - A_R$
240	-0.53×10^{-4}
260	4.00×10^{-4}
280	1.05×10^{-4}
300	-0.31×10^{-4}

  What are the concentrations of DNA and RNA in the mixture?
  d. If the DNA and RNA are hydrolyzed to give the individual nucleotides, will the molar ellipticity increase, decrease, or remain the same?

4.3. A common feature of many DNA binding proteins is the "leucine zipper." It is two similar sequences of 30–35 amino acids containing multiple leucines. Each sequence forms a right-handed α-helix, and the two helices wrap around each other to form a left-handed super helix. A number of studies have been carried out of this "coil-coil" structure with model peptides. In one such study the circular dichroism spectum at $0°$C had large positive peak at 195 nm and negative minima at 208 and 222 nm. This was interpreted as being the spectrum of a completely helical structure.

  a. At $0°$C, the molar ellipticity at 222 nm is $-33,000$ deg cm^2 dmol^{-1}. At $80°$C, the molar ellipticity at 222 nm is essentially zero. Interpret this result.

  b. At $55°$C, the molar ellipticity at 222 nm is $-16,000$ deg cm^2 dmol^{-1}. What percentage of the peptide is α-helical at this temperature?

  c. For some model systems, it is found that the temperature at which the ellipticity approaches zero increases as the concentration of model peptide increases. How would you explain this result?

4.4. Estrogen receptors are ligand-activated transcription factors that mediate the effects of female sex hormone on DNA transcription. The interaction of estrogen receptors with a specific DNA fragment that binds the receptor and estradiol has been studied with circular dichroism [N. Greenfield, V. Vijayanathan, T. J. Thomas, M. A. Gallo, and T. Thomas, *Biochemistry* **40**, 6646 (2001)]. The circular dichroism spectrum was analyzed to show that the receptor is approximately 75% α-helical, 3% β-sheet, 10% turns, and 12% random coil. The following data were obtained for the ellipticity at 222 nm in the presence and absence of the DNA and estradiol.

Ligand	$[\theta]_{222}$, deg cm^2 dmol^{-1}
None	−25,000
5 μM DNA	−32,000
5 μM estradiol	−22,000

a. Assume that the specific rotation at 222 nm is due only to the α-helix and that the receptor is saturated with the ligand. Indicate what is happening to the structure of the receptor, and calculate the percentage of helix when the ligand is bound.

b. When the temperature is raised, the ellipticity at 222 nm approaches zero. Explain what is happening to the structure of the receptor.

c. The temperature at which the difference between the initial and final ellipticity has reached half of its value, T_m, is 38.0°C in the absence of ligands. It is 43.6°C and 46.1°C in the presence of 5 μM estradiol and 5 μM of the specific binding DNA, respectively. Explain these results.

d. In the presence of 750 nM estradiol, T_m is 40.8°C. Explain this result, and estimate the dissociation constant for the binding of estradiol to its receptor protein with the assumption that the concentration of estradiol is much greater than the receptor concentration. At 40.8° C, the ratio of denatured protein to native protein was determined to be 1.20 in the absence of ligands. (This is a hypothetical result as this result was not reported in the publication.)

4.5. Adenosine monophosphate has an extinction coefficient of about 15,000 cm^{-1}M^{-1} at 260 nm, 0°C. The value of $\varepsilon_L - \varepsilon_R$ at 260 nm is about 2.00 cm^{-1} M^{-1}. The corresponding value of $\varepsilon_L - \varepsilon_R$ for a polymer of polyrA is −17.0 cm^{-1}M^{-1} at 0°C, and its magnitude decreases greatly as the temperature is raised. (The monomer concentration was used in calculating this number.)

a. Explain the reason for the large difference in the differential extinction coefficient between AMP and polyrA.

b. Why does the magnitude of the differential extinction coefficient decrease as the temperature is increased?

c. Calculate the observed ellipticity in degrees for a 10^{-4} M solution of AMP and for a 10^{-4} M solution of polyrAMP (monomer concentration) at 260 nm, 0°C.

d. Calculate the absorbance of a 10^{-4} M solution of AMP. Would the absorbance of the polyrA at the same concentration (0°C) be greater than, less than, or the same as the ATP solution?

4.6. The system for transport of mannitol across the *Escherichia coli* membrane involves a membrane-bound enzyme that couples phosphorylation of the sugar to its translocation through the membrane. Fluorescence lifetime and anisotropy measurements have been carried out of the tryptophans in this enzyme dissolved in detergent micelles. [D. Dijkstra, J. Broos, A. J. W. G. Visser, A. van Hoek, and G. T. Robillard, *Biochemistry* **36**, 4860 (1997)]. The fluorescence anisotropy decay of one of these tryptophans has two components. The faster component, with a rotational correlation time of about 1 ns, is due to local motion of the tryptophan. The decay of anisotropy for the slower component can be approximated by the following data.

Anisotropy	Time (ns)
0.290	0.00
0.245	5.00
0.208	10.00
0.149	20.00
0.126	25.00
0.106	30.00
0.090	35.00

a. Calculate the rotational correlation time.

b. The expected correlation time for the known molecular weight of the macromolecule is 120–140 ns. How do you explain the difference between the calculated and observed correlation time?

c. When the protein is phosphorylated, the rotational correlation time increases to 51 ns. What might be the cause of this change?

CHAPTER 5

VIBRATIONS IN MACROMOLECULES

INTRODUCTION

Thus far we have considered only spectral phenomena in the visible and ultra-violet regions of the spectrum. In this spectral region, the interaction between light and molecules causes transitions between energy levels of the electrons and alters the populations of electronic energy levels. Molecules also have characteristic vibrational motions that are influenced by much longer wave-lengths, typically in the far red, or infrared. Since the energy associated with a given wavelength, λ, is hc/λ, this means the difference in energy between levels is much smaller than for the electronic energy scaffold. (Recall that h is Planck's constant, and c is the speed of light.) For macromolecules the number of different vibrations is approximately 3N, where N is the number of atoms in the molecule. For small molecules the number of translational and rotational degrees of freedom must be subtracted from this number, but this is a small correction for macromolecules.

In principle, some type of coupling might be expected when transitions occur between electronic and vibration energy levels. However, this is not the case because electronic transitions occur much more rapidly than the time scale of nuclear motions—nuclei are quite sluggish compared to electrons (10^{-16} s vs. 10^{-13} s). Thus the nuclei can be assumed to be stationary during an electronic transition. This is called the Franck-Condon principle. The Franck-Condon principle is why electronic transitions between vibrational energy levels of different electronic energy states can be drawn as straight lines (Fig. 3-11).

A detailed analysis of all of the vibrational modes of freedom of a macromolecule is not possible, but a few general conclusions bear mention. First, the characteristic vibration of each degree of freedom is a combination of the motions of many different bonds. These characteristic motions are called *normal modes* of vibration. The nature of these normal mode vibrations can be calculated very precisely for small molecules, and the characteristics of some of the normal modes can be associated with similar normal modes in macromolecules. Second, in some limiting cases these normal modes are dominated by the movement of a single or a restricted number of chemical bonds. For example, the vibrations of C–H bonds, double bonds between C, and double bonds between C and O are each associated with similar spectral characteristics in many molecules. Third, the energy level distribution for each normal mode can be approximated by the relationship

$$E = \left(v + \frac{1}{2}\right)h\upsilon \tag{5-1}$$

where E is the energy and v is the vibrational quantum number which has integral values from 0 to very large values. This relationship is derived directly from quantum mechanics assuming that the vibration is characterized as a simple harmonic oscillator, essentially a spring moving back and forth. Note that in the lowest energy state, v = 0, the normal mode still has an intrinsic energy, $(\frac{1}{2})h\upsilon$. This is called the zero-point energy and is possessed by all molecules, even at the hypothetical temperature of absolute zero. The zero point energy is a manifestation of the uncertainty principle (Eq. 1-4): the energy cannot be zero because this implies the positions would be known exactly.

The potential energy, U, for a harmonic oscillator is shown in Figure 5-1, together with the energy levels inside the potential well. For a simple harmonic oscillator,

$$U = \left(\frac{1}{2}\right)kx^2 \tag{5-2}$$

In this equation, k is a constant and x is the vibrational coordinate. If the oscillator is thought of as a one-dimensional spring, the x coordinate is the coordinate that the spring moves back and forth along, and k is a measure of how stiff the spring is. For a single bond the x coordinate is the distance along the bond, and the origin of the coordinate system is the equilibrium bond distance. The characteristic frequency of the bond motion is

$$v = (1/2\pi)(k/\mu)^{1/2} \tag{5-3}$$

where μ is the reduced mass of the system. For a complex normal mode, the reduced mass is a weighted average of the masses: for our purposes, an exact calculation of the reduced mass is not necessary. For a C–H bond, the reduced

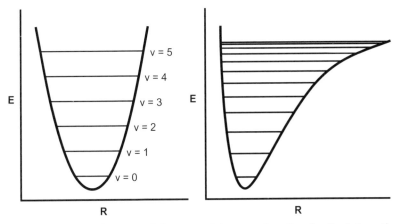

Figure 5-1. Schematic diagrams of the potential energy function for the interaction of two atoms in a diatomic molecule. A harmonic oscillator potential is shown on the left and a more realistic intermolecular potential on the right. R is the distance between the atoms, and the energy levels are shown within the wells. For a potential energy function obeying Eq. 5-2, the abscissa for the potential on the left is the x coordinate with x = 0 at the minimum in the curve. The actual number of energy levels is much greater than shown, but the even spacing of the energy levels in the harmonic oscillator potential is apparent, as is the decreasing space between energy levels as v increases for the more realistic potential energy function. When the energy reaches the top of the well, the diatomic molecule dissociates into atoms.

mass can be approximated as the mass of the hydrogen atom. A useful tool for identifying the nature of a given vibrational mode is to substitute deuterium for hydrogen. According to Eq. 5-3, the ratio of the characteristic frequency of the vibration is

$$v_H/v_D = (m_D/m_H)^{1/2} = \sqrt{2} \tag{5-4}$$

This substitution of deuterium for hydrogen is very useful for determining if the motion of a hydrogen atom is the dominant factor in a given vibrational mode.

For real molecules, the vibrational motion deviates from this simple harmonic model at very high energies, and the energy levels become more closely spaced at the top of the potential well. The approximate potential energy function for a vibrational coordinate of a real molecule (anharmonic oscillator) is included in Figure 5-1. For a diatomic molecule, the abscissa is the internuclear distance, whereas for larger molecules it is a combination of internuclear distances associated with the normal mode vibration. Quantum mechanical calculations can be carried out with anharmonic potential energy functions to provide direct correlation of theory with experimental findings. Note that near the bottom of the potential well, harmonic and anharmonic oscillators behave quite similarily so that the harmonic oscillator is a good model at room temperature.

INFRARED SPECTROSCOPY

The difference in energy between different vibration energy levels is about 100 J/mole. This corresponds to light in the infrared region, $\lambda \sim 1$ mm. Wave numbers, $\bar{v}$, are usually used when discussing vibration spectra, rather than wavelength. The wave number is simply the reciprocal of the wavelength. Quantum mechanical calculations indicate that there are selection rules that govern the transitions between vibrational energy levels. If absorption of light is to occur, the vibration must cause a change in the electric dipole moment. Recall that a dipole moment is simply a measure of the balance of charges within a molecule. For example, HCl has a permanent dipole moment because H has a partial positive charge and Cl has a partial negative charge. On the other hand, H_2 does not have a permanent dipole moment because no net imbalance in charge is present. From a practical standpoint, it is important to note that water has a dipole moment and absorbs light in the infrared very strongly. Consequently, it is very difficult to measure the infrared absorption of macromolecules in water as the absorption-emission of the molecule is significantly obscured by that of water. Sometimes a combination of H_2O and D_2O can be used to partially circumvent this problem, but it severely restricts the use of infrared spectroscopy for biological systems. In some cases, dried and/or hydrated films are used, although this is far from the conditions in the milieu of biology.

Two experimental methods are available for measuring infrared spectra. One is essentially the same as the visible ultraviolet spectrometer previously discussed. Infrared light is passed through the sample, and the absorption is measured. The only difference is the light source, typically a glowing wire, and the detector, typically a thermocouple. This method has been largely supplanted by Fourier transfer methods. These methods were briefly described in Chapter 1. Basically with Fourier transform infrared measurements, a beam of light is split in two, with only half of the light going through the sample. The difference in phase of the two waves creates constructive and/or destructive interference and is a measure of the sample absorption. The waves are rapidly scanned over a specific wavelength region of the spectra, and multiple scans are averaged to create the final spectrum. This method is more sensitive than the conventional dispersion spectrometer.

RAMAN SPECTROSCOPY

Another method exists for studying transitions between vibration energy levels that uses a concept not yet discussed. In addition to absorbing light, samples also scatter light. The amount of scattered light is a maximum 90° to the direction of the incident beam. Most of the scattered light is at the same frequency as that of the incident light. This is called Rayleigh scattering. At the molecular level, the electric field of the light perturbs the electron distribution, but no transitions between energy levels occur so that the molecule immediately returns to its unperturbed state. This scattering is inversely proportional to the fourth power of the wavelength so the scattering is much greater at shorter wavelengths. This is essentially why the sky is blue, the shorter wavelengths of the visible spectrum (blue) are scattered more than the longer wavelengths (red). Rayleigh scattering is observed at all wavelengths. The intensity of the scattered light is related to the *polarizability* of the molecule, that is, to how easily electrical charges can be shifted within the molecule to make it more polar. A small number of the molecules return to a different vibrational energy level after scattering. The vibrational energy level can be either higher or lower than the initial state. As a result of this change in energy level, some of the scattered light will be at a slightly lower or higher frequency than the incident light. This is called *Raman* scattering, after the Indian scientist who discovered the phenomena.

A typical setup for measuring Raman scattering is shown in Figure 5-2. A very intense light source is needed to observe Raman scattering because only a very small amount of the scattered light displays a change in frequency. The advent of lasers has permitted this to be done routinely with visible light and in the ultraviolet with rather expensive lasers. Prior to lasers, very large arc lamps were utilized. In addition, very high concentrations are required. However, Raman scattering does not require the light to be at a wavelength comparable to the energy of vibrational transitions since it is a *scattering* phenomenon, as contrasted to absorption spectroscopy. The Rayleigh line is very intense, but much less intense scattering can be detected at a lower frequency than the incident light. This is because the incident light was used to promote the molecule to a higher vibrational energy level. These are called Stokes lines. An even smaller fraction of the scattered light occurs at a higher frequency than the incident light because energy is added to the incident light by the movement of the molecule to a lower vibrational energy level. These are called anti-Stokes lines. Most molecules are found in their ground vibrational energy level at room temperature so that the observation of anti-Stokes lines is rare.

Raman spectroscopy has two major advantages over infrared spectroscopy for studying transitions between vibrational energy levels. First, a permanent dipole moment is not required. It is only necessary for the polarizability of the molecule to change between different vibrational energy levels. Second, visible light can be used rather than infrared light so that Raman spectra can be readily obtained in water. They can also be obtained in crystals and films. As previously indicated, the primary disadvantages are that because the intensity of the Raman lines is very weak, intense light sources and high concentrations of the molecule of interest are needed. Raman and infrared spectroscopy should be regarded as complimentary. Since infrared spectroscopy is depend-

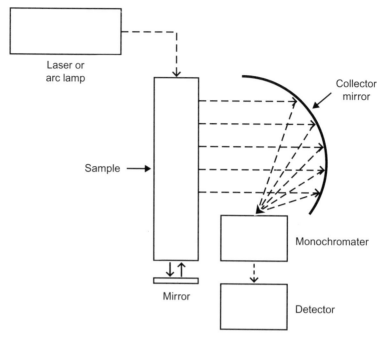

Figure 5-2. Schematic diagram of an apparatus used for Raman spectroscopy. The mirror at the end of the sample cell is to put the light through the sample a second time, effectively increasing the path length of the cell, and the collector mirror is designed to collect as much of the scattered light 90° to the source light beam as possible.

ent on the permanent dipole moment and Raman spectroscopy on the polarizability, usually (but not always) a vibrational transition is observed either in the infrared or in Raman scattering, but not in both.

Thus far in our discussion, the implicit assumption is that the wavelength of the exciting light for Raman spectroscopy is not near an absorption band. If the wavelength coincides with the absorption band for an electronic transition, a large increase occurs in the intensity of the Raman spectrum. Basically this is related to the fact that an electronic absorption band is the superposition of many different vibrational modes. This can sometimes be seen directly in the electronic absorption spectrum by fine structure in the peaks. Essentially the alteration of the population of vibrational energy levels within the electronic absorption band is responsible for the enhanced intensity of the Raman spectrum. Determining the spectrum within the electronic absorption band is called *resonance Raman spectroscopy*. The two-fold advantage of resonance Raman spectroscopy is that lower concentrations can be used and the spectrum is simplified because only the intensified lines are observed. At the present time, resonance Raman measurements are made primarily in the visible region of the spectrum. In biological systems this means chromophores such as hemes and retinal must be present.

STRUCTURE DETERMINATION WITH VIBRATIONAL SPECTROSCOPY

Infrared spectroscopy has been a useful tool for the determination of the structure of organic molecules for many years. This is because specific types of bonds and/or chemical groups have characteristic vibrational frequencies. Some of these group frequencies are given in Table 5.1. Infrared spectra are often referred to as "fingerprints" for the molecule and large compilations of data are available. In the case of biological molecules, changes in the group frequencies can be used to derive information about the secondary structure of the molecules.

The carbonyl of the amide bond in proteins is particularly useful for the determination of secondary structure (2, 3). The stretching normal mode, amide I mode, of the carbonyl has been shown to have a specific frequency associated with α-helices, β-sheets, and other characteristic structures. (Strictly speaking, this normal mode is not just the stretching of the carbonyl. It also involves some bending of the C-N-H angle.) This was ascertained by the study of model peptides for which precise measurements and theoretical calculation of normal modes could be carried out. The approximate wave numbers corresponding to the three common structures found in proteins are: α-helix, $1650\,cm^{-1}$; β-sheet, 1632 and $1685\,cm^{-1}$; and random coil, $1658\,cm^{-1}$.

For proteins a more empirical approach has been adopted. The most successful approach uses known structures to calibrate the vibrational frequency measurements. The vibrational spectrum of the amide bond for a protein is complex because of the many amide bonds present in multiple environments. Nevertheless, the spectrum can be deconvoluted to provide information about the amount and types of secondary structures present. The assumption usually made is that the observed vibrational frequency is a linear combination of the frequencies associated with the various secondary structures that are present, with each specific frequency weighted by the percent of a given structure. Some typical results are shown in Table 5.2 and compared with the known structure of the protein and CD estimates of secondary structure (4). Raman spectroscopy has proven particularly useful for studying secondary structure of proteins since water solutions can be used, and the amide I mode has a spectral band that is well isolated from other protein bands ($1630–1700\,cm^{-1}$). The amide I band is the most common vibrational frequency used as an indicator of secondary structure (cf. 5), but other bands and their relation to structure have been identified (3).

The structure of nucleic acids also can be investigated with vibrational spectroscopy. The base vibrations and the groups involved in hydrogen bonding are particularly sensitive to the secondary structure of nucleic acids. Most of the work with nucleic acids has involved Raman spectroscopy (3). For

TABLE 5-1. Group frequencies in the infrared region[a]

Chemical group	Frequency (cm^{-1})
—CH$_3$	1460
—CH$_2$—	2930
	2860
	1470
C—H	3300
—C—C—	1165
—C=O	1730
—C—H (in CH$_3$)	2960
	2870
—C—H (in CHO)	2870
	2720
H	3060
—CN	2250
—O—O—	1200–1100
—OH	3600
—NH$_2$	3400
=CH$_2$	3030
—SH	2580
—C=N—	1600
C—Cl	725
C=S	1100

[a] Reproduced with permission from D. Sheehan, Physical Biochemistry, John Wiley and Sons, 2000, p. 98

example, the melting of DNA and RNA structures can be readily followed, and the specific vibrational frequencies provide molecular details about the melting process. The different types of helices formed by DNA can be distinguished. Again, standards are established with known structures and then used to determine the structures of unknown samples, that is, the amount of various types of structures present.

ESONANCE RAMAN SPECTROSCOPY

me proteins have been extensively studied with resonance Raman spec-
scopy (6). They are very prevalent in nature and have a very intense
orbance in the visible region of the spectrum due to the porphyrin ring
cture. Excellent Raman spectra can be obtained at very low concentra-
s, often in the micromolar range. The vibrational spectra obtained are char-
ristic of the porphyrin ring structure. However, the highest frequency
al modes, 1350–1650 cm^{-1}, are very sensitive to the state of the Fe atom
s bound to the porphyrin. Thus, these frequencies can establish the spin
and coordination states of the Fe in both of its oxidation states. The
g of ligands to the Fe and the distortion of the porphyrin skeleton also
detected. As an example, the high-frequency region of the spectrum for
and CO-myoglobin is shown in Figure 5-3 (6). Myoglobin is a protein
used for oxygen transport in some organisms, a function carried out by
lobin in humans. The frequency shifts in the vibration spectra between
wo forms of myoglobin can be readily discerned.

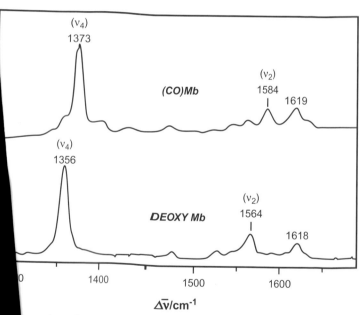

mparison between resonance Raman spectra for carbonmonoxy-
Mb] and deoxy-myoglobin at room temperature. The excitation was in
egion corresponding to the heme absorption. Note that the two species
distinguished by their Raman spectra. Reprinted from T. Spiro and
wicz, Resonance Raman Spectroscopy of Metalloproteins, *Meth.*
6 (1995). © 1995, with permission from Elsevier.

TABLE 5-2. Protein Secondary Structure Determined by Infrared (I Dichroism (CD) Spectra and X-Ray Crystallography[a]

Protein	Secondary Structure (			
	α-Helix	β-Sheet	Turn	R
Hemoglobin	78	12	10	
	87	0	7	
	68–75	1–4	15–20	
Myoglobin	85	7	8	
	85	0	8	
	67–86	0–13	0–6	
Lysozyme	40	19	27	
	45	19	23	
	29–45	11–39	8–26	
Cytochrome c (oxidized)	42	21	25	
	48	10	17	
	27–46	0–9	15–2{	
α-chymotrypsin	9	47	30	
	8	50	27	
	8–15	10–53	2–	
Trypsin	9	44	38	
	9	56	24	
Ribonuclease A	15	40	3	
	23	46	2	
	12–30	21–44		
Alcohol dehydrogenase	18	45		
	29	40		
Concanavalin A	8	58		
	3	60		
	3–25	41–49		
Immunoglobin G	3	64		
	3	67		
Major histocompatability complex antigen A2	17	41		
	20	42		
	8–13	74–7⁻		
β_2-macroglobulin	6	52		
	0	48		
	0	59		

[a] Reproduced with permission from A. Dong, P. Hu (1990). © 1990 American Chemical Society.
[b] The band due to random structure appears as a to be separated from α-helix structure. The rando in the α-helix value.

R

He
tro
abs
stru
tion
acte
norn
that
state
bindi
can b
deoxy
that is
hemog
these t

Intensity

13(

Figure 5-3. C
myglobin [(CO
the wavelength
can be readily
R. S. Czernusze
Enzymol. **246**, 41

tonated in the DHFR-H$_2$folate complex at neutral pH, but it was protonated int the DHFR-H$_2$folate-NADP$^+$ complex. A pH titration indicated the pK of N5 is 6.5 in this complex, as compared to 2.6 when the substrate is not bound to the enzyme. In the actual reaction, NADPH is present rather than NADP$^+$ so that the implicit assumption is that the complex studied is a good model for the catalytic reaction. Thus, N5 can be protonated much more readily in the enzyme-NADP$^+$ complex.

From the point of view of understanding enzyme catalysis, this result means that the population of the N5 protonated substrate is four orders of magnitude larger in the environment of the enzyme. The positive charge on the N5 would also make C6 more positive and would presumably make the hydride transfer reaction much faster than with unprotonated N5. Much of the catalytic effect of the enzyme, therefore, apparently is due to providing an environment that stabilizes the protonated substrate prior to the hydride transfer. A structural explanation for this stabilization is not obvious. The N5 is in a hydrophobic region of the protein, and no negative charges are conveniently close that might stabilize the protonated species. However, conformational changes of the protein have been established as part of the overall mechanism of action of the enzyme, and these changes may be the source of the stabilization. In any event, Raman spectroscopy has provided unique insight into the catalytic process.

Although vibration spectroscopy has not been used as extensively in biological systems as ultraviolet-visible absorption and fluorescence spectroscopy, it can sometimes provide unique and important information.

REFERENCES

1. K. Sauer, ed., *Meth. Enzymol.* **246** (1995).

2. F. Siebert, *Meth. Enzymol.* **246**, 501 (1995).

3. W. L. Peticolas, *Meth. Enzymol.* **246**, 389 (1995).

4. A. Dong, P. Huang, and W. S. Caughey, *Biochemistry* **29**, 3303 (1990).

5. W. K. Surewicz, H. H. Mantsch, and D. Chapman, *Biochemistry* **32**, 389 (1993).

6. T. Spiro and R. S. Czernuszewicz, *Meth. Enzymol.* **246**, 416 (1995).

7. D. L. Rousseau and J. M. Friedman, in *Biological Application of Raman Spectroscopy* (T. G. Spiro, ed.), Vol. 3, p. 133, Wiley, New York, 1988.

8. M. Nagai, M. Aki, R. Li, Y. Jin, H. Sakai, S. Nagatomo, and T. Kitagawa, *Biochemistry* **39**, 13093 (2000).

9. H. Deng and R. Callender, in *Infrared and Raman Spectroscopy of Biological Materials* (H.-U. Gremlich and B. Yan eds.), Marcel-Dekker, Inc., New York, p. 477–514 (2001).

10. Y.-Q. Chen, J. Kraut, R. E. Blakley, and R. Callender, *Biochemistry* **33**, 7021 (1994).

PROBLEMS

5.1. The zero point energies for two different vibrational manifolds are 9.55×10^3 and 1.19×10^4 Joules/mole.

   a. Calculate the frequency of light emitted for a transition between the first energy level and the zero point energy for each of these manifolds.

   b. In terms of the harmonic oscillator model, which of these manifolds has the "stiffer" spring.

   c. For both cases, the radiation associated with a transition between energy level 20 and 19 occurs at a longer wavelength (smaller wave number) than the radiation associated with a transition between the energy level 2 and 1. How do you explain this?

5.2. From infrared studies of model proteins, the wave numbers for the amide II band (due to N–H deformation in the peptide bond) are 1540–1550 cm^{-1} for the α-helix, 1520–1525 cm^{-1} for β-sheet, and <1520 cm^{-1} for "random coils".

   a. Calculate the zero point energy for these three vibration energy levels.

   b. Explain why the three fundamental wave numbers differ in terms of the protein structure.

   c. For polyglutamic acid, the amide II band is at a wave number of about 1545 cm^{-1} at low pH (pH 4). As the pH increases (>pH 9), the wave number decreases to below 1520 cm^{-1}. Explain this result.

CHAPTER 6

PRINCIPLES OF NUCLEAR MAGNETIC RESONANCE AND ELECTRON SPIN RESONANCE

INTRODUCTION

In this chapter, we consider the interaction of molecules with radiation when molecules are placed into a strong magnetic field. The fundamental properties of atoms that are important for this discussion are the nuclear spin, for nuclear magnetic resonance (NMR), and the electron spin, for electron spin resonance (ESR) or, equivalently, electron paramagnetic resonance (EPR). Strictly speaking, the concept of spin can be rigorously defined only by the use of quantum mechanics. However, we will use a semiclassical approach in which spin in the nucleus or in an electron can be represented as a charge moving in a circular path. This movement creates a magnetic dipole that can be thought of as a bar magnet. In the absence of a magnetic field, the magnetic dipole is oriented randomly, and only one energy level is associated with the electron or nucleus. But in the presence of a magnetic field, the magnetic dipoles (or bar magnets) tend to be oriented either in the direction of the field or opposed to it, thus creating multiple energy states. Application of quantum mechanics to this situation indicates that the orientation of the magnetic dipole and the energy states are quantized with characteristic quantum numbers.

For electrons, the spin quantum number, S, is 1/2, and the two spin states are +1/2 and −1/2, represented as the familiar arrows pointed up or down. Neutrons and protons also have a spin quantum number of 1/2 so that the nucleus

Spectroscopy for the Biological Sciences, by Gordon G. Hammes
Copyright © 2005 John Wiley & Sons, Inc.

of an atom has a characteristic spin quantum number, I. Simple rules exist for determining the nuclear spin quantum number. Nuclei with an even mass number and even charge number have no nuclear spin (I = 0). Nuclei with an odd mass number have a half integral spin (I = 1/2, 3/2, 5/2 etc.). Finally, nuclei with an even mass number and odd charge number have integral spin (I = 1, 2, etc.).

The number of energy levels associated with spin is determined by the spin quantum number. For an electron, the number of energy levels is 2S + 1, or 2. In the absence of a magnetic field, no distinction can be made between these two quantum states: They have the same energy, that is, they are degenerate. In the presence of a magnetic field, however, the alignment of the magnetic moments with and against the magnetic field creates two distinct energy levels. The hydrogen nucleus also has a spin quantum number, I, of 1/2, with two possible orientations of its magnetic dipole (or magnetic moment) in a magnetic field and two energy levels (2I + 1). On the other hand, ^{23}Na has I = 3/2 and four energy levels in the presence of a magnetic field. In general, the quantum states in the presence of a magnetic field are characterized by quantum numbers ranging from I to –I in integral steps. Thus, for sodium, these quantum numbers are 3/2, 1/2, –1/2, and –3/2. Although the average orientation of the nuclear spin state of a proton is either aligned with the magnetic field or against it, the conservation of angular momentum requires the magnetic moment associated with the orientation to rotate about the direction of the field. This is analogous to a top spinning on its axis and rotating around a vertical line due to gravity.

A quantum mechanical treatment of nuclear spins in a magnetic field of a strength H provides an explicit equation for the energy levels, E:

$$E = -g_N \beta_N H M_I = -(h/2\pi)\gamma H m_I \qquad (6\text{-}1)$$

In this equation, β_N is the nuclear magneton and is a universal constant calculated from the properties of nuclei: $\beta_N = 5.051 \times 10^{-27}$ Joules/Tesla. The nuclear g factor, g_N, is a constant, but it is different for each atom, and m_I is the spin quantum number characterizing the orientation of the magnetic moment in the magnetic field (I to –I in integral steps). This equation also defines the gyromagnetic ratio, γ, which is a frequently used constant. Values of the nuclear spin quantum mumber, I, and γ for some nuclei of biological interest are presented in Table 6.1.

The dependence of the energy on magnetic field for atoms with a total nuclear spin quantum number of 1/2 is shown in Figure 6-1. The difference in energy between the quantized levels increases as the magnetic field increases. Also shown is the precession of the magnetic moment about the field direction for the two possible orientations.

The energy difference between the energy levels can be calculated, again with the assistance of quantum mechanics, by the requirement that the change

TABLE 6-1. Magnetic Properties of Selected Nuclei

Isotope	Spin	$10^7\gamma$ $(T^{-1}s^{-1})$	Natural Abundance
^1H	1/2	26.75	99.98
$^2H(D)$	1	4.11	0.0156
^{13}C	1/2	6.73	1.108
^{14}N	1	1.93	99.63
^{15}N	1/2	−2.75	0.37
^{19}F	1/2	25.18	100.0
^{31}P	1/2	10.84	100.0
^{17}O	5/2	−3.63	0.037

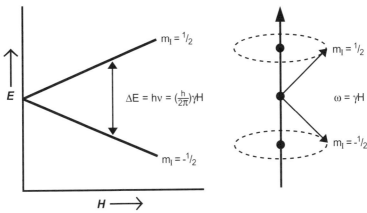

Figure 6-1. On the left, a schematic plot of the energy versus the magnetic field is shown for a nuclear or electron spin with a quantum number of 1/2. The frequency of the radiation emitted/absorbed for a transition between the two energy states created by the magnetic field is dependent on the strength of the magnetic field (Eq. 6-2). On the right, the magnetic dipole associated with the nuclear spin is shown precessing around the direction of the magnetic field at an angular frequency ω. The two orientations, up and down, correspond to the two energy levels in the diagram on the left.

in quantum number for allowed transitions between energy levels is 1. In this case

$$\Delta E = g_N\beta_N H = (h/2\pi)\gamma H \qquad (6\text{-}2)$$

With this equation we can calculate the frequency, υ, associated with transitions between energy levels since $\Delta E = h\upsilon$. The magnetic field of modern instruments varies from about 7 to 19 Tesla. For a magnetic field of 11.75 Tesla, the frequency associated with transitions between the energy levels of protons is 500 MHz. This is in the radio frequency range. The actual energy difference between levels is quite small, only 3.4×10^{-25} J/proton. Because the energy difference is small the actual population difference between energy levels is also very small at room temperature: the ratio of populations in the two states is $\exp(-\triangle E/kT) = 0.99993$ for an 11.75 Tesla magnetic field at 37°C. For this same magnetic field, the resonant frequency for ^{13}C is 130 MHz.

The quantum mechanical calculation of the energy levels for an unpaired electron in a magnetic field is quite similar to that for nuclear spin, except that the electron always has spin 1/2. The energy difference between levels is

$$\Delta E = g_S \beta_S H \tag{6-3}$$

where $g_S = 2.0023$ and $\beta_S = 9.274 \times 10^{-24}$ Joules/Tesla for electrons. Because the Bohr magneton is much larger for an electron than a proton, the energy levels are further apart, and the resonance energy frequency is much larger. (β_S and β_N differ by the ratio of the mass of the proton to the mass of the electron, 1836.) For a 1 Tesla field, the resonant frequency is 28,000 MHz = 28 GHz. This frequency is in the microwave region, and quite different experimental techniques are required for ESR and NMR.

Finally, we return to the precession of the magnetic moment about the direction of the magnetic field, as depicted in Figure 6-1. The angular precession frequency, that is, how fast the magnetic moment is rotating about the vertical line, is $\omega = 2\pi\upsilon = \gamma H$. This is called the *Larmor* frequency. The visualization of rotating magnetic moments is useful when considering the effects of changing magnetic fields on the nuclear spins.

We first discuss NMR in some detail because it is extensively used in biology. Although ESR has been used to obtain important information about biological systems, it is less extensively used and will receive relatively brief consideration.

NMR SPECTROMETERS

The first NMR spectrometers placed a sample in a fixed magnetic field and applied a fixed radiofrequency by means of a coil perpendicular to the field direction. The magnetic field was varied by a coil until resonance was achieved, with the absorption being detected by a third coil. This is analogous to the methods used in visible and ultraviolet spectroscopy. Although the magnetic field applied to the sample is uniform, the actual magnetic field at the nucleus is dependent on a number of environmental factors so that the field must be scanned to find the energy absorption (resonance) condition. We will return to this matter a bit later. A typical spectrum would display the absorption of energy versus frequency, as shown schematically in Figure 6-2.

The absorption of radiation is difficult to detect because the populations of the energy states are very similar, as discussed above. In practice, this means that relatively high concentrations of the species being observed must be present. The sensitivity of detection depends on the characteristics of the

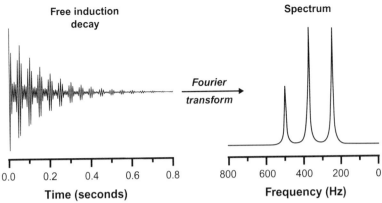

Figure 6-2. The free induction decay of the signal from nuclei in a magnetic field is shown (*left*) after a short pulse of radiofreqency radiation is applied to the sample. A Fourier transform of the free induction decay gives the familiar nmr spectrum on the right where the absorption is plotted versus the frequency. Copyright by Professor T. G. Oas, Duke University. Reproduced with permission.

nucleus being observed and its natural abundance. The sensitivity also is enhanced by increasing the energy difference between the two states, which is the reason that instruments with larger magnetic fields are being developed continuously. The proton, ^{1}H, provides the best sensitivity, and its natural abundance is 99.98%. Consequently, the most extensive measurements have been carried out with proton NMR. Although the natural abundance of ^{31}P is essentially 100%, the sensitivity of detection relative to protons is only about 6%. Probably the second most studied atom with NMR is ^{13}C. Its natural abundance is only about 1%, but its prevalence in biological compounds is very extensive. Furthermore, its abundance in molecules of interest can be enhanced through synthesis or bacterial growth with ^{13}C enriched compounds. Regrettably, the most abundant isotope of carbon, ^{12}C, does not have a nuclear spin. A variety of other isotopes have been studied with NMR. Most notable for biologists are ^{14}N, ^{15}N, and ^{19}F, which can often be substituted for H in substances of biological interest.

As discussed in Chapter 1, the frequency dependence of the absorption can be transformed into a time dependence and vice versa through Fourier transforms. The most common method of obtaining an NMR spectrum today is to apply a timed radio frequency pulse and then watch the nuclei return to their equilibrium configurations. The time dependence of the return to equilibrium or *free induction decay* (FID) can be transformed from a time-dependent signal into a frequency spectrum, as shown schematically in Figure 6-2. Multiple FIDs can be combined to produce an average FID that has less noise than a single FID. Consequently, Fourier transform instruments are considerably more sensitive than continuous wave instruments so that lower concentrations can be used. The timing and nature of the pulses can be quite complex, but this does not alter the basic concept underlying Fourier transform methods.

CHEMICAL SHIFTS

Thus far we have discussed NMR as though the nuclei were isolated in the magnetic field, but the utility of NMR derives from the interaction of the nuclei with surrounding electrons and other nuclei in the molecule. The external magnetic field interacts with the electrons to induce magnetic moments in the electrons that usually oppose the external field. Consequently, the magnetic field at the nucleus is usually lower than the external field. This shielding effect of the electrons can be incorporated into the standard equations by noting that the field at the nucleus, H, in a static magnetic field, H_0, is

$$H = H_0(1 - \sigma) \tag{6-4}$$

where σ is the shielding constant, typically about 10^{-5}. The value of σ can be positive or negative, depending on whether the magnetic field from the electrons aligns against or with the external magnetic field.

The shielding effect is directly proportional to the strength of the external field, but the same relative change in resonance frequency is observed, regardless of the external field strength. This shielding effect, therefore, can be expressed as a relative change in frequency with respect to a standard, thereby rendering it independent of the external field. These frequency changes are called *chemical shifts*, δ, and are given in units of parts per million (ppm). The chemical shift can be written as

$$\delta = \frac{v - v_{ref}}{v_{ref}} 10^6 = (\sigma - \sigma_{ref})10^6 \tag{6-5}$$

where v is the frequency of the nucleus, v_{ref} is the frequency of a standard compound, and σ_{ref} is the shielding constant of the standard compound. The most common reference compound for protons is tetramethylsilane, but it is insoluble in water so trimethylsilylpropionate-d_4 is usually used in aqueous media (cf. reference 1 for a discussion of chemical shift references). In principle, the chemical shift depends on the orientation of the sample with respect to the magnetic field. In liquids, this is generally not a problem because molecules are rapidly tumbling and sampling all possible orientations. For large molecules that tumble relatively slowly, however, the resonances can become so broad that the spectrum is obscured, and in solids special conditions are required to obtain high-resolution spectra.

The electron density at the nucleus is often the dominant factor in determining the chemical shift. A high electron density creates a large shielding, and the applied magnetic field must be increased to get resonance: this results in an upfield shift and a decrease in the magnitude of δ because reference com-

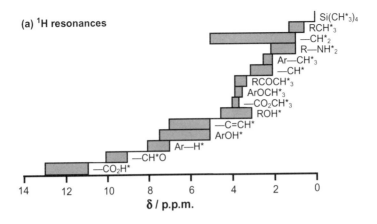

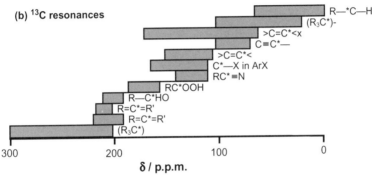

Figure 6-3. Range of typical chemical shifts for ^1H and ^{13}C resonances. P. W. Atkins, *Physical Chemistry*, 3rd edition, W. H. Freeman, New York, NY, 1986, p. 489. © 1978, 1982, 1986 by Peter W. Atkins. Used with permission of W. H. Freeman and Company.

pounds are generally highly shielded. Conversely, a low electron density at the nucleus causes a downfield shift and increase in δ. The range of chemical shifts for ^1H and ^{13}C in various compounds are shown in Figure 6-3.

In principle, the chemical shifts of nuclei in proteins should provide information about the protein structure. However, chemical shifts alone are not sufficient to determine protein structure. The average chemical shifts of various nuclei for amino acids in denatured proteins are given in Table 6.2. Note that they are quite similar for all of the amino acids. When secondary structures such as α-helices or β-sheets are present, changes in chemical shifts occur so that the amounts of various secondary structures present can be inferred from the NMR spectra.

Some of the largest changes in chemical shifts in proteins and nucleic acids are observed for aromatic rings of nucleotides, tyrosine, phenylalanine, and

TABLE 6-2. Average Chemical Shifts of Random Coil Amino Acids (ppm)

Amino Acid	α-^{1}H	Amide-^{1}H	α-^{13}C	Carbonyl-^{13}C	Amide-^{15}N
Ala	4.33	8.15	52.2	177.6	122.5
Cys	4.54	8.23	56.8	174.6	118.0
Asp	4.71	8.37	53.9	176.8	120.6
Glu	4.33	8.36	56.3	176.6	121.3
Phe	4.63	8.30	57.9	175.9	120.9
Gly	3.96	8.29	45.0	173.6	108.9
His	4.60	8.28	55.5	174.9	119.1
Ile	4.17	8.21	61.2	176.5	123.2
Lys	4.33	8.25	56.4	176.5	121.5
Leu	4.32	8.23	55.0	176.9	121.8
Met	4.48	8.29	55.2	176.3	120.5
Asn	4.74	8.38	52.7	175.6	119.5
Pro	4.42	—	63.0	176.0	128.1
Gln	4.33	8.27	56.0	175.6	120.3
Arg	4.35	8.27	56.0	176.6	120.8
Ser	4.47	8.31	58.1	174.4	116.7
Thr	4.35	8.24	62.0	174.8	114.2
Val	4.12	6.19	62.2	176.0	121.1
Trp	4.66	8.18	57.6	173.6	120.5
Tyr	4.55	8.28	58.0	175.9	122.0

Reproduced with permission from D. S. Wishart and B. D. Sykes, Chemical Shifts as a Tool for Structure Determination, *Meth. Enyzmol.* **239**, 363 (1994). © 1994, with permission of Elsevier.

tryptophan. These shifts are due to the interaction of the external magnetic field with the delocalized electrons of the aromatic ring. Nuclei above or below the ring usually have decreased chemical shifts, whereas those near the edges have increased chemical shifts. These effects are called *ring currents* and can cause unusually large chemical shifts. These chemical shifts can provide information about the structure of macromolecules and about alterations in structure due to changes in the environment such as temperature and the addition of various chemical agents.

SPIN-SPIN SPLITTING

The chemical shift is caused by interactions between the nucleus and nearby electrons. A conceptually different interaction is transmitted between nearby nuclei by intervening electrons participating in chemical bonds. Basically, the spin state of a neighboring nucleus alters the shielding a nucleus experiences. This effect is smaller than typical chemical shifts and is called *spin-spin splitting* and is often referred to as scalar coupling. Unlike chemical shifts, the magnitude of spin-spin coupling is independent of the magnitude of the applied magnetic field.

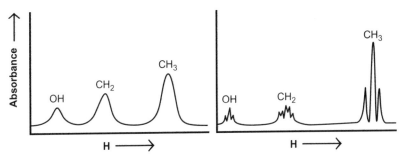

Figure 6-4. Schematic representation of the nmr spectrum of dry ethanol at low (*left*) and high (*right*) resolution. The ratio of the areas under the peaks is 3:2:1. The octet expected for the methylene protons (*right*) is not shown as the splitting of the four peaks is very small.

This effect is most easily understood by considering the proton NMR spectra of ethanol shown in Figure 6-4. The low-resolution spectrum (left) shows the three peaks that might be expected on the basis of our discussion of chemical shifts, namely CH_3 protons, CH_2 protons, and the OH proton. The ratio of the areas under the resonances is proportional to the number of protons in each chemical environment, 3:2:1. The high-resolution structure (right) shows that these three resonances are multiplets of peaks due to spin-spin splitting. The number of peaks within a multiplet is determined by the number of spin orientations of neighboring nuclei. In this case, the spins of the two hydrogens on the methylene (CH_2) carbon can have four possible arrangements of orientations

$$\uparrow\uparrow \quad \underline{\uparrow\downarrow} \quad \underline{\downarrow\uparrow} \quad \downarrow\downarrow$$

The middle two arrangements are equivalent so that the methyl proton resonances are split by the neighboring CH_2 into three resonances with relative areas under the peak of 1:2:1. The OH proton, also next to the methylene carbon, is split into three peaks. Similarly, the methyl proton can have three different orientations, with the two underlined groups being equivalent:

$$\uparrow\uparrow\uparrow \quad \underline{\uparrow\uparrow\downarrow} \quad \underline{\uparrow\downarrow\uparrow} \quad \underline{\downarrow\uparrow\uparrow} \quad \underline{\uparrow\downarrow\downarrow} \quad \underline{\downarrow\uparrow\downarrow} \quad \underline{\downarrow\downarrow\uparrow} \quad \downarrow\downarrow\downarrow$$

The methylene resonance, therefore, is split into four peaks, with area ratios of 1:3:3:1. The hydroxyl proton will split these four peaks further into an octet, but this detail is difficult to see in the spectrum. In order for scalar coupling to occur, the two nuclei must have distinct resonance peaks. Thus the methyl protons will not split each other's resonances.

Spin-spin splitting is of special importance in considering the NMR spectra of proteins because coupling between the α-C proton and N proton of the amide bound occurs. Scalar coupling can also occur between different nuclei, for example, a proton and ^{15}N. The magnitude of the spin-spin splitting is called the coupling constant and is designated by J. The number of covalent bonds separating the nuclei in question is often appended as a prior superscript and a post subscript designates the atoms involved. For example, the coupling constant for the amide proton and the α-C proton would be written as $^3J_{HN-H\alpha}$.

The spin-spin coupling constant for protons on adjacent atoms varies considerably, from about 0 to 10 Hz. This variation is due to different torsional angles between the protons as defined below:

$$\theta=0° \qquad \theta=180°$$

If the torsional angle is 90°, the coupling is close to zero, whereas when it is 0° or 180°, it is about 10 Hz. If free rotation about the bond occurs, an average coupling constant is obtained. The dependence of the coupling constant on the torsional angle is given by the Karplus equation:

$$J = A + B\cos\theta + C\cos^2\theta \qquad (6\text{-}6)$$

The constants A, B, and C can be calculated or established from measurements with molecules having known dihedral angles. If free rotation about the bond occurs, the coupling constant will be some average of the possible dihedral angles. However, if the bond is constrained, such as in a peptide linkage that has double bond character or in a folded macromolecular structure, the calculated dihedral angel can provide useful information about the structure. This is especially true when used in conjunction with other information.

This simple picture of spin-spin splitting is not rigorous. It provides a useful and adequate explanation for relatively simple situations but breaks down when considering multidimensional NMR, which is discussed later.

RELAXATION TIMES

Thus far we have not dealt explicitly with the time scale for nuclear spin transitions. If we think of a pulsed NMR experiment, we can envisage nuclear spins being oriented in a specific direction by the pulse, with the precession of spins about the direction of the field. When the pulse is turned off, the nuclei will come to equilibrium with regard to their environment. The rate at which a particular nuclear spin returns to equilibrium depends on interactions with other nuclear spins, and this in turn will depend on the fluctuating fields experienced as the molecules tumble. This mode of relaxation involves a change of energy between the spin systems and their environment. The return of a spin population in a magnetic field to its equilibrium population follows first-order kinetics, and the reciprocal of the first-order rate constant is called the *spin-lattice relaxation time*, T_1. Typically it is in the range of tenths of a second to seconds for protons.

Simply put, the value of T_1 depends on the interactions of a nuclear spin with its neighbors, including the solvent, and how fast the molecule rotates. The molecular rotation is characterized by a rotational correlation time that is basically a measure of the rotational diffusion constant of a molecule. Measurements of T_1 can, in fact, be used to determine rotational diffusion constants of rigid molecules. In some cases, chemical exchange of nuclei in different environments can contribute to T_1 although this is somewhat unusual. In general, anything that gives rise to magnetic fluctuations in the environment can contribute to this relaxation, for example, unpaired electrons or dissolved oxygen.

A second mode of nuclear spin relaxation is possible that does not involve the exchange of energy of the magnetic moment with its environment. In terms of the picture of a nuclear spin being oriented in a field and precessing about the field direction, the spin-lattice relaxation time can be thought of as characterizing the return of the spin orientations to their equilibrium positions. The *spin-spin relaxation time*, T_2, on the other hand, is associated with the rate of precession about the field direction. Basically, T_2 is a measure of alterations in the precession frequency during nuclear spin relaxation. This alteration is different for different nuclear spins so that the rates of precession change with respect to each other. This is called a loss of phase coherence. This can be viewed as an exchange of energy within the spin system. It does not change the net population of the excited states.

The dominant factor determining T_2 is the rate of molecular tumbling. The molecular tumbling effect is quantitatively different for T_2 and T_1. Because T_2 is generally much shorter than T_1 in liquids, the line widths of spectra are determined by T_2, $1/T_2 = \pi \upsilon_{1/2}$ where $\upsilon_{1/2}$ is the peak width at one-half of its maximum value. As previously stated, rapidly tumbling molecules produce relatively sharp lines whereas slowly tumbling molecules have relatively broad lines.

Chemical reactions provide one of the most interesting examples of how T_2 can be altered. Consider the simple chemical reaction

$$A \rightleftharpoons B \tag{6-7}$$

Assume that a specific proton has a different chemical shift in A and B. (This means they have a different Larmor frequency.) If the chemical exchange rate is very slow compared to the difference in chemical shifts (i.e., smaller than the difference in Larmor frequencies), two distinct peaks will be seen in the NMR spectrum, as shown in Figure 6-5 (top). At the other extreme, if the chemical exchange rate is very fast relative to the difference in the chemical shift frequencies, states A and B will interconvert many times during the NMR experiment, and the two frequencies are effectively averaged. Consequently,

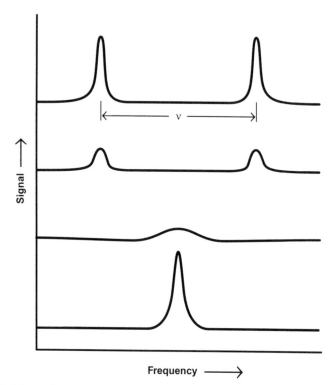

Figure 6-5. Schematic drawing of the effect of chemical exchange on NMR spectra. In the top spectrum, resonances from two protons are shown. In this case, the rate of exchange between the two environments is much less than υ, or $1/\tau_A \ll \upsilon$, whereas in the lowest spectrum, $1/\tau_A \gg \upsilon$, so that only a single resonance is seen that is the average of the positions in the upper spectrum. The two spectra in the middle represent the cases where $1/\tau_A \sim \upsilon$, with $1/\tau_A$ becoming progressively larger from the top spectrum to the bottom spectrum.

only a sharp single line will be seen, located at the average of the two lines weighted by the relative populations of A and B (Fig. 6-5, bottom). In the intermediate cases, where the rate of the chemical reaction is comparable to the difference in chemical shift frequencies, the two lines will broaden, coalesce into a single broad peak, and finally sharpen, as shown in Figure 6-5. For a single line, the spin-spin relaxation time can be written as

$$1/T_2 = 1/T_{2A} + 1/\tau_A \qquad (6-8)$$

where T_{2A} is the relaxation time of the A state without significant chemical exchange and τ_A is the relaxation time for the chemical reaction. We will not delve into the details, but the line shape can be quantitatively analyzed to determine the rate constants for the chemical reaction. The time scale of the reactions that can be studied is determined by the chemical shift difference between the two states. For example, if the difference is 100 Hz, then rates of chemical reactions in the range of $1/100 = 10^{-2}$ s can be studied. Reactions such as the rate of exchange of hydrogens between proteins or amino acids and water have been investigated with this method.

MULTIDIMENSIONAL NMR

Thus far we have considered what is now called one-dimensional NMR, namely determining the spectrum for a specific nucleus by scanning the magnetic field or by analysis of the frequencies associated with the free induction decay following a frequency pulse. One-dimensional analysis has been invaluable in determining the structures of small molecules and can also be used to obtain information about macromolecules. However, in order to get definitive structural information about macromolecules, multidimensional NMR is necessary. The genesis of this field was in the early 1970s and its vigorous evolution continues to this day. The underlying principle of multidimensional NMR is to find "cross-peaks" that link two resonances. This linkage can be either through space or through a small number of chemical bonds. Finding these cross-peaks allows the spatial relationships to be determined between the nuclei responsible for the two resonances. These connections between resonances are sometimes called coherence pathways.

A detailed presentation of multinuclear NMR is beyond the scope of this text. However, the concepts can be understood by considering two-dimensional NMR in a qualitative manner. Two-dimensional NMR can be viewed as the assembly of one-dimensional spectra in an array. The experiments consist of four stages, illustrated in Figure 6-6. In the preparation phase a frequency pulse is applied to the system. An evolution phase of length t_1 then

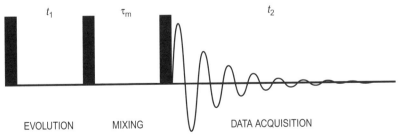

Figure 6-6. Schematic representation of a two-dimensional NMR experiment in which a radiofrequency pulse (black bar) is applied to the sample initially. After a period t_1, a second pulse is applied, followed by a mixing time, τ_m. The data are acquired after a final pulse. The number of pulses and the various times depend on the type of experiment being carried out.

occurs and is followed by a mixing period ending in a second pulse. (A pulse may also be applied between the evolution and mixing phases, depending on the specific two-dimensional experiment being carried out.) Finally, a free induction decay occurs for a time t_2 in the data acquisition phase. In the actual experiment, t_1 is varied incrementally and the signal is collected for the period t_2, ultimately giving a large number of time points. This experiment is repeated many times and signal averaged. The nature of the pulses will depend on the specific experiment. Qualitatively, the pulses flip the spins (magnetic moments) in the field that is created, and the end of the mixing phase flips them again. In the evolution phase, the magnetic moments associated with the nuclear spin will partially return to equilibrium at a rate that depends on T_1 and T_2. The return to equilibrium also occurs in the collection phase. In contrast, one-dimensional NMR uses a single pulse, followed by free induction decay and data collection.

The design of these experiments can be quite tricky, as it depends on what connections between spin states is being probed and what the relaxation times are. The representation of the result is done by carrying out a Fourier transform on both the t_1 and t_2 data sets and converting them into a two-dimensional plot of υ_1 versus υ_2 with the third dimension being the amplitude of any resonance peaks that are observed. A very simple illustration is given in Figure 6-7 for two interacting spin systems. Peaks that occur on the diagonal have the same frequency in both dimensions and correspond to the one-dimensional spectrum, whereas those occurring off the diagonal represent cases where different spin systems interact during the mixing period (coherence transfer). Contours are usually used to indicate the amplitudes of the resonance peaks, rather than a third dimension. The interactions (coherence transfer) between nuclear spins can be either homonuclear (same nuclei) or heteronuclear (different nuclei).

One of the first two-dimensional experiments carried out was COSY (COrrelated SpectroscopY). This experiment identifies pairs of nuclei that are linked by scalar coupling (spin-spin splitting connectivities). The interaction between these nuclei occurs during the relaxation taking place in the mixing phase and results in cross-peaks in the spectrum. Scalar coupling (off diagonal peaks) between nuclei means they are within three covalent bonds. A typical COSY spectrum is shown in Figure 6-8 for λ cro repressor, a DNA binding protein that regulates phage development.

One of the most important two-dimensional NMR experiments is Nuclear Overhauser Effect SpectroscopY (NOESY). This is a through space interaction that takes place because of the interactions between the magnetic dipoles of two nuclear spins. In terms of our previous discussion, this primarily involves the T_1 mode of relaxation and is coupled with rotational motion. The NOE is the NMR equivalent of fluorescence resonance energy transfer discussed in Chapter 3. In fact, an NOE can be observed in one dimension. If a sufficiently large magnetic field is applied at the resonance condition of a given nucleus, the spin system becomes saturated, that is, the ground and excited

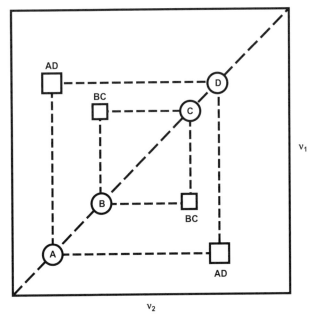

Figure 6-7. Schematic representation of a two-dimensional NMR spectrum for two spins. The frequency axes labeled υ_1 and υ_2 are the positions of the resonances for the two different spins. The diagonal (circles) corresponds to the one-dimensional spectrum. If coherence transfer occurs during the mixing period, off diagonal resonances (squares) will be seen. In this example, coherence transfer occurs between nuclei with resonances at A and D and between nuclei with resonances at B and C.

states become equally populated so that no more energy can be absorbed. If the magnetic moment of this nucleus is sufficiently close in space to another magnetic moment, energy can be transferred. This perturbs the intensity of the resonance of the nucleus to which energy is transferred and decreases the resonance intensity of the nucleus that was initially irradiated. The interaction between magnetic dipoles varies as the inverse sixth power of the distance between the two dipoles so that only nuclear spin systems that are very close to each other give rise to NOEs. In practice, this means distances of 5 Å or less.

The NOESY experiment is quite similar to the two-dimensional experiment described previously, except that a pulse is applied at both the beginning and end of the mixing period, with the time of the mixing period being constant. An example of a NOESY spectrum is given in Figure 6-9 where proton-proton NOEs for a complex of DNA and an antibiotic, distamycin A, are presented. Because of the distance dependence of this effect, the distance between spins can be estimated from the cross-peaks. However, because only the distance, not the direction, is derived from these measurements, many distances must be determined to arrive at a unique structure.

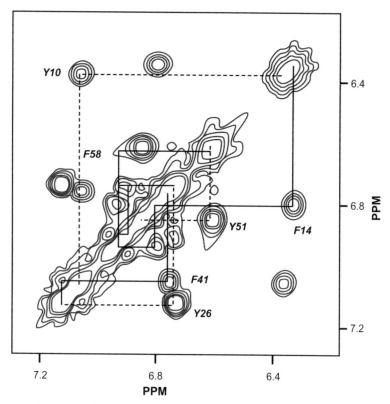

Figure 6-8. Example of a COSY spectrum for the protons of the aromatic spin systems in λ cro repressor protein. Tyrosine rings are connected with a dashed line and phenylalanine rings by a solid line. The diagonal and cross-peaks can be easily seen. Reprinted in part with permission from P. L. Weber, D. F. Wemmer, and B. R. Reid, *Biochemistry* **24**, 4553 (1985). © 1985 by American Chemical Society.

One of the most useful multidimensional spectra is HSQC (Heteronuclear Single Quantum Correlation), an example of which is shown in Figure 6-10 for a protein involved in proton transport across a membrane (subunit c of ATP synthase from *E. coli*). This spectrum uses both ^{1}H and ^{15}N and selectively detects only pairs of covalently attached nuclei. Each spot in the contour map represents such a pair, with the position on the horizontal axis representing the proton resonance frequency and the position on the vertical axis representing the resonance frequency of the nitrogen nucleus. Every amino acid, except for proline, has a backbone amide so essentially every residue is represented in this spectrum. Side chains containing amides will give rise to additional resonances. This gives direct information about which nitrogen is coupled to which hydrogen. Furthermore, the resonances are usually quite well

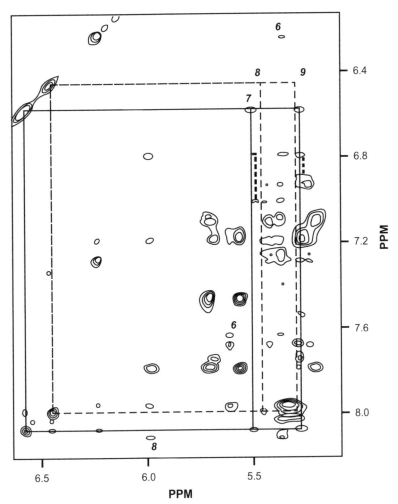

Figure 6-9. Example of a NOESY spectrum for a complex of distamycin A and DNA. Aromatic C6H resonances of adenine and guanine and C2H resonances of adenine are shown along the vertical axis. The Cl'H resonances are shown along the horizontal axis. Sets of sequential connectivities are denoted by dotted, dashed, and solid lines. Reprinted in part with permission from J. G. Pelton and D. F. Wemmer, *Biochemistry* **27**, 8088 (1988). © 1988 by American Chemical Society.

resolved. The HSQC spectrum provides information that is useful for assigning the observed resonances to specific amino acid residues.

Extending NMR to dimensions greater than two involves similar concepts. Multiple pulses and even field gradients are used. These dimensions can be in terms of different nuclei and/or combining two-dimensional experiments. The most common nuclei studied in addition to ^{1}H are ^{13}C and ^{15}N. In these cases,

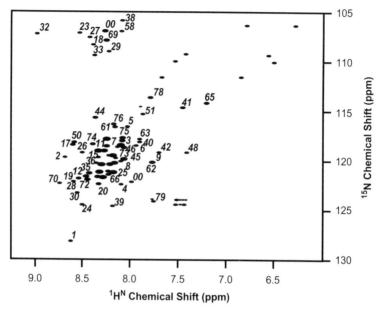

Figure 6-10. Example of an HSQC spectrum. All 76 backbone amide cross-peaks are shown for the c subunit protein of ATP synthase from *E. coli*. This membrane-bound protein is involved in proton transport across the membrane. Reprinted in part from M. F. Girvin, V. K. Rastogi, F. Abildgaard, J. L. Markley, and R H. Fillingame, *Biochemistry* **37**, 8817 (1998). © 1998 by American Chemical Society.

proteins and nucleic acids enriched in the nucleus of interest can be used to enhance the sensitivity.

Transforming the NMR results for a macromolecule into a structure is straightforward but not easy. First, the resonances have to be assigned to specific nuclei within the structure. This can be done by analysis of scalar coupling and sequential NOEs. The three-dimensional structure is derived from the distances determined from NOE experiments and dihedral angle information determined from spin-spin splitting and the Karplus equation (Eq. 6-6). In practice, the information gleaned from the NMR spectra provides hundreds of constraints on the structure. A number of computer programs have been written that take the constraints and convert them into a family of structures, usually very similar, consistent with the constraints. These programs involve sophisticated theory as well as data analysis to arrive at final structures. NMR spectroscopy is a very powerful tool for determining protein structures in their biologically active conformations. As proteins increase in size, their rate of rotation slows down, and NMR spectra broaden. Although the size of macromolecules whose structure can be determined with NMR increases year by year, most of the structures to date are for molecules with a molecular weight less than 20,000.

MAGNETIC RESONANCE IMAGING

One of the most remarkable advances in diagnostic medicine has been the evolution of magnetic resonance imaging (MRI). Although x rays readily distinguish hard objects, such as bones, they do not distinguish soft tissue structure. MRI, on the other hand provides excellent images of tissues and is able to distinguish between various types of tissue. Protons in water are the primary nucleus used for detection although applications with ^{13}C, ^{31}P, and ^{19}F have been developed. The principle underlying MRI is the use of a magnetic field gradient. Since the resonance frequency is directly proportional to the magnetic field, the frequency of the resonance will depend on its location in the magnetic field. The intensity of the absorption is dependent on how many protons are present. If a linear magnetic field gradient is applied, the position of the resonance will change, also linearly, as the field is varied. Thus a plot of the amplitude of the resonance versus frequency is equivalent to a plot of the integrated number of protons versus distance. A series of cross-sections can be obtained by rotating the sample in the field, or by moving the field around the sample. These cross-sections can then be reconstructed to give a three-dimensional image.

In soft tissue, the amount of water varies for different tissues, so the density of protons varies. In addition, the various tissues are characterized by significantly different T_1 values. The difference in proton density can be shown in reconstructions by varying the darkness of the shading. This can be seen in Figure 6-11 where the MRI image of an adult human brain is shown. In addition to being able to distinguish various soft tissues, MRI is noninvasive, as contrasted to x rays or injections of foreign substances, including radioactive isotopes, required for other imaging techniques.

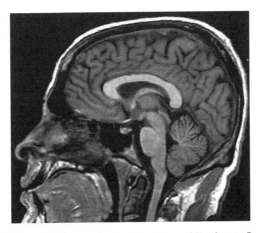

Figure 6-11. MRI of an adult human brain. Courtesy of Professor Scott Huettel, Duke University Brain Imaging and Analysis Center. Reproduced with permission.

ELECTRON SPIN RESONANCE

We now return briefly to ESR. Unlike NMR, ESR is not observed for most materials. This is because electrons are usually paired and consequently have no net magnetic moment. However, free radicals and other paramagnetic substances have unpaired electrons that give rise to ESR spectra. Because the frequencies associated with transitions between the energy levels of unpaired electrons in a magnetic field are in the microwave region, special techniques are required for placement of the sample in the magnetic field. Conceptually the experiment is the same as for NMR. The magnetic field is varied until resonance is found.

The usefulness of ESR in biological systems arises from the interactions between nuclear spins and the electron spin. This gives rise to "hyperfine structure" in the spectrum. For example, if a neighboring nucleus has a nuclear spin of 1/2, it will have two orientations in the field and will split each of the two energy states of the electron into two, as shown in Figure 6-12. Because of quantum mechanical selection rules, only two transitions between these four energy states are allowed. If the nuclear spin is 1, six energy levels are created and three transitions between the levels are allowed. The hyperfine structure can provide information about the environment of the paramagnetic species.

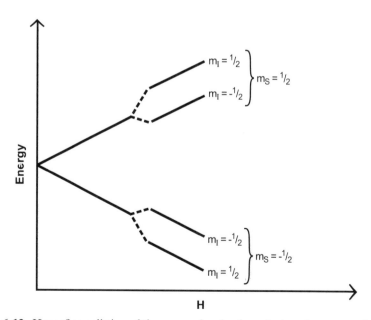

Figure 6-12. Hyperfine splitting of the energy levels of an electron in a magnetic field, H, by a nuclear spin. The electron has a spin quantum number, m_S, of $\pm 1/2$, and the nucleus has a spin quantum number, m_I, of $\pm 1/2$.

As with NMR, the sharpness of the spectrum provides information about the rotational mobility of the paramagnetic species. However, the time scale for rotation is much different than NMR because the frequencies are much higher. For ESR, rotational correlation times of approximately 10^{-9} s influence the line width whereas for NMR the time scale is about 10^{-6} s. For reasons we will not dwell on here, the spectra are usually presented as the derivative of the amplitude versus the field (Fig. 6-13).

One of the important developments in the application of ESR to biological systems was the synthesis of stable free radicals that could be reacted with macromolecules and membranes, both noncovalently and covalently (2). The most common element of these spin labels is the nitroxide free radical (Fig. 6-14). Because the ^{14}N nucleus, with a nuclear spin of 1, is next to the free radical, three bands are seen in the spectrum. The derivative of the resonances is three up and down peaks. This is illustrated in Figure 6-14 where ESR spectra are shown for nitroxides under various conditions of rotational mobility. The effect of molecular motion on the spectra of the free radical is clearly illustrated. As these spectra illustrate, spin labels can provide information about the rotational mobility of the macromolecule to which they are bound. However, it should be noted that the probe mobility can be due to macromolecule rota-

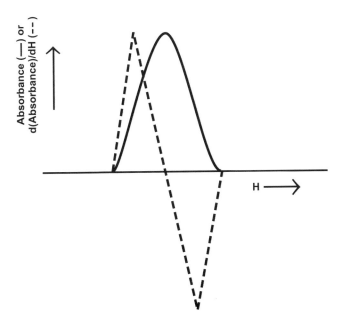

Figure 6-13. Schematic representation of a derivative spectrum. An absorbance peak is shown as a solid line, and its derivative is shown as a dashed line. ESR spectra are usually presented as derivatives of the absorption.

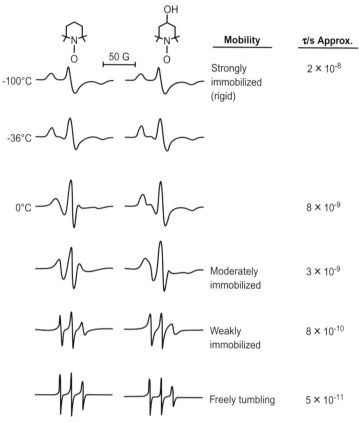

Figure 6-14. ESR spectra of spin labels whose structures are shown at the top of the figure. The effect of viscosity on the line shapes and rotational correlation times, τ, is shown. Reproduced with permission from R. A. Dwek, *Nuclear Magnetic Resonance in Biochemistry*, Clarendon Press, Oxford, England, 1973, p. 289. Adapted from data in P. Jost, A. S. Waggoner, and O. H. Griffith, Spin Labeling and Membrane Structure, in *Structure and Function of Biological Membranes* (L. Rothfield, ed.), Academic Press, New York, 1971, p. 83. © 1971, with permission of Elsevier.

tion, segmental motion of the macromolecule, or simply rotation of the probe in the site to which it is bound. The interpretation of what motion is being observed is not always straightforward.

ESR spectra can also provide information about the polarity of the spin label environment. The extent of splitting of the hyperfine structure depends on the effective dielectric constant of the environment. Thus, for example, the splitting is quite different in a biological membrane or in solution. When bound to a membrane, the splitting will be different for a spin label close to a polar headgroup and for a spin label buried in the hydrocarbon chains.

This chapter is a relatively brief introduction to NMR and ESR. Many texts are available that provide more detailed descriptions of NMR(cf. 3–7) and ESR (8). Unfortunately, most of these are not easy reading and require a working knowledge of quantum mechanics. Applications of these techniques to biological systems are presented in the next chapter.

REFERENCES

1. D. S. Wishart and B. D. Sykes, *Meth. Enzymol.* **239**, 363 (1994).
2. S. Ohnishi and H. M. McConnell, *J. Am. Chem. Soc.* **87**, 2293 (1965).
3. I. Tinoco, Jr., K, Sauer, J. C. Wang, and J. D. Puglisi, *Physical Chemistry: Principles and Applications to the Biological Sciences*, 4th edition, Prentice Hall, Englewood Cliffs, NJ, 2001.
4. K. Wuthrich, *Accts. Chem. Res.* **22**, 36 (1989).
5. T. L. James and N. J. Oppenheimer, eds., *Meth. Enzymol.* **239** (1994).
6. J. Cavenaugh, W. J. Fairbother, A. G. Palmer III, and N. J. Skelton, *Protein NMR Spectroscopy*, Academic Press, San Diego, CA, 1996.
7. T. L. James, V. Dötsch, and U. Schmitz eds., *Meth. Enzymol*, **238** and **239** (2001).
8. J. A. Weil, J. R. Bolton, and J. E. Wertz, *Electron Paramagnetic Resonance: Elementary Theory and Practical Applications*, Wiley-Interscience, New York, 1994.

PROBLEMS

6.1. Deduce the structure of the compounds below from their schematic NMR spectra. Indicate which protons are assigned to each resonance.

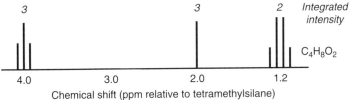

Chemical shift (ppm relative to tetramethylsilane)

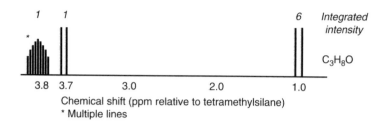

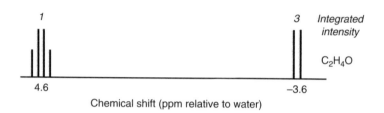

6.2. Nucleotide phosphates are important biological molecules. Sketch the ^{31}P (spin 1/2) NMR spectra of a nucleotide monophosphate, diphosphate, and triphosphate. Order the resonances in terms of chemical shift, with zero on the right-hand side. The phosphates are designated as α, β, and γ with α being closest to the sugar and γ furthest away. For the triphosphate, the γ position has the lowest electron density and the β position has the highest.

6.3. The structure of L-leucine in D_2O is

$$
\begin{array}{ccc}
 & CH_3 & H \\
 & | & | \\
H-C_\gamma-C_\beta H_2-C_\alpha-COO^- \\
 & | & | \\
 & CH_3 & ND_3^+
\end{array}
$$

a. Sketch the one-dimensional proton NMR spectrum. The approximate chemical shifts are 1.0 for the methyl groups, 1.4 for the γ hydrogen, 1.5 for the β hydrogens, and 3.3 for the α hydrogen. Assume that the methyl hydrogens and β hydrogens are not resolved, that is, all of the hydrogens in each of the two classes have a single resonance peak. Indicate the integrated intensity of the resonances for each resonance.

b. Sketch the proton COESY spectrum of L-leucine, indicating the coupling that should give rise to off-diagonal peaks.

6.4. Two common structures found in proteins are the α-helix and β-sheet (parallel and antiparallel), as discussed in Chapter 2, The approximate distances in Å between protons for the two structures are given below:

	α-Helix	β-Sheet
amide H—amide H	2.8	4.2
α-carbon H—amide H	3.5	2.2
α-carbon H—(amide H)$_{i+3}$	3.4	>5

The subscript i + 3 means the α-carbon is i + 3 residues away from the amide proton. Indicate how the measurement of NOE's could be used to distinguish these structures by considering the expected relative intensities of the NOEs for the three classes of protons in the table.

6.5. a. Use Eq. 6-3 to construct a plot of the resonant frequency for an electron as a function of the field strength from 0 to 4 Tesla.

b. Show schematically the splitting expected in the energy levels if the unpaired electron is located next to a nuclear spin of 3/2.

6.6. An RNA structure containing uracil and adenine was studied by NMR. The structures of these two bases are given below.

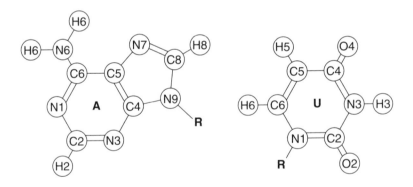

a. In a $^2J_{NN}$ COSY type experiment with ^{15}N-enriched RNA, scalar couplings were observed between N3 of uridine and N1 of adenine [A. J. Dingley and S. Grzesiek, *J. Am. Chem. Soc.* **120**, 8293 (1998)]. In addition, in a NOESY type experiment, an NOE was observed between H3 of uridine and H2 of adenine. Postulate a structure for the hydrogen-bonded dimer formed between uridine and adenine.

b. As the pH of the solution is lowered, the chemical shift of adenine N1 shifts downfield, and at sufficiently low pH (<4), the scalar coupling and NOE disappeared. Explain this result. (This experiment was not actually done, but this is the expected result.)

c. With free adenosine, this same chemical shift is observed, but it occurs at much lower pH values. Explain this result.

6.7. The spectrum of pure methanol is shown in *a* below. When HCl is added, the spectrum is that shown in *b* [Z. Luz, D. Gill, and S. Meiboom, *J. Chem. Phys.* **30**, 1540 (1959)].

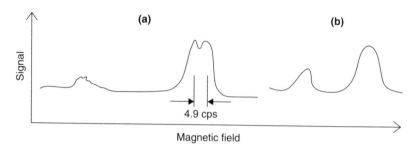

a. Explain this result.

b. The rate of the proton exchange reaction

$$CH_3OH_2^+ + CH_3OH \rightarrow CH_3OH + CH_3OH_2^+$$

can be written as $R = k(CH_3OH_2^+)(CH_3OH)$ where the rate constant k has a value of about $10^8 \, M^{-1} s^{-1}$. The chemical relaxation time, τ, for this reaction is $\tau = 1/[k(CH_3OH_2^+)]$. If the above reaction is responsible for the doublet in the spectrum becoming a single peak, estimate the concentration of HCl at which the doublet disappears. Is this consistent with the autoprotolysis constant methanol: $(CH_3O^-)(CH_3OH_2^+) = 2 \times 10^{-17} \, M^2$?

CHAPTER 7

APPLICATIONS OF MAGNETIC RESONANCE TO BIOLOGY

INTRODUCTION

Magnetic resonance has been widely applied to biological systems so that the selection of examples to illustrate the information that can be obtained is necessarily quite arbitrary. We first consider some examples of structural determinations utilizing NMR. As previously indicated, the advantage of NMR is that structures in solution can be obtained and variation of these structures in different environments can be readily assessed. The primary drawback of NMR is that only relatively small structures can be easily determined. NMR can provide important information about biological systems other than macromolecule structures, as will be illustrated. In the case of ESR, spin labels and paramagnetic ions can provide unique information, especially in unusual environments such as membranes.

REGULATION OF DNA TRANSCRIPTION

The regulation of transcription is of central importance in biology. We have previously discussed the use of CD to study zinc fingers, a structure of special importance for the interaction of transcription factors with DNA (Chapter 4). In this section, we consider the structure of cyclic-AMP response element binding protein (CBP). It is a large transcriptional adapter protein that medi-

Spectroscopy for the Biological Sciences, by Gordon G. Hammes
Copyright © 2005 John Wiley & Sons, Inc.

ates transcription responses to intra- and extracellular signals (cf. 1, 2). This class of proteins has been implicated in the regulation of cell growth, transformation, and differentiation. Defects in this regulator are involved in a multitude of human diseases. The CBP interacts with a variety of transcription factors and other components of transcription regulation to mediate cellular activities.

CBP itself is a very large protein, more than 2000 amino acids, but contains a number of distinct structural and functional domains. Three putative zinc-binding domains (zinc fingers) were identified by sequence homology. The CH3 domain was selected for study because not only is it a representative zinc finger structure, it also binds known transcription factors (3). Constructs of varying length were examined, but the final studies were done on an amino acid sequence that was 88 residues long. It contains 13 cysteine and 5 histidine residues. Nine of the cysteine residues were coordinated to Zn^{2+}. This was ascertained from the chemical shifts of $^{13}C^{\beta}$ of cysteine. The cysteines that were not coordinated to Zn^{2+} had chemical shifts of 26.5–27.1 ppm, whereas the cysteines bound to Zn^{2+} had chemical shifts of 29.2–30.7. The three Zn^{2+}-coordinated histidines also displayed a down field shift of 8–9 ppm for a ^{13}C ring resonance of the imidazole and an upfield shift for an ^{15}N resonance. Thus, the sites of metal coordination could be determined from one-dimensional NMR. Three Zn^{2+} ions are bound/CH3 domain.

If Zn^{2+} is absent, the NMR spectra indicated a disordered structure. However, in the presence of the metal a well-defined structure formed. This can be seen in Figure 7-1, where the HSQC spectrum is shown for the protein with three bound Zn^{2+} atoms. Many additional cross-peaks can be seen in the presence of Zn^{2+}. The structure was solved using three-dimensional heteronuclear spectra of uniformly labeled ^{15}N and ^{15}N, ^{13}C protein. The structure determination involved the fitting of NOESY and dihedral angle constraints. In addition an energy minimization was carried out to arrive at the final structure(s). Energy minimization is a theoretical calculation of the minimum free energy for the structure making use of empirical equations for the intramolecular interactions. The final structure is shown in Figure 7-2 (see color plates) in the usual format, namely the best fit superposition of multiple structures, in this case 20. In essence, this representation provides a measure of the uncertainty in the final structure. The cysteines and imidazoles are also shown as ball-and-stick representations to indicate how the Zn^{2+} is bound to the protein. Table 7.1 is a brief summary of some of the NMR constraints used to determine the structure and the deviations from ideal covalent geometry.

The structure contains four α-helices and three histidine(cysteine)$_3$ Zn^{2+} binding motifs. The helices are tightly packed to form a hydrophobic core. Two of the metal binding regions are quite similar, but the structure of the third is noticeably different. Although detailed data were not presented, preliminary results suggest that the structure of the CH1 domain is quite similar to that of CH3. Surprisingly, the structures of these zinc fingers are different than others

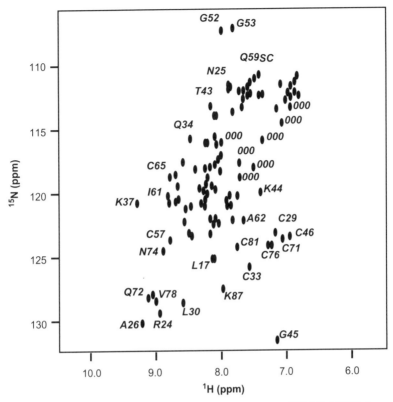

Figure 7-1. The 600 MHz proton-nitrogen HSQC spectrum of TAZ2 (CH3) domain, a portion of the c-AMP response element binding protein, with three bound Zn^{2+}. Many cross-peaks can be seen. Some of the amino acid assignments for the resonances are indicated. Reprinted from R. N. De Guzman, H. Y. Liu, M. Martinez-Yamout, H. J. Dyson, and P. E. Wright, Solution Structure of the TAZ2 (CH3) Domain of the Transcriptional Adaptor Protein CBP, *J. Mol. Biol.* **303**, 243 (2000). © 2000, with permission from Elsevier.

that have been determined, presumably because of unique steric constraints within the structure.

As mentioned at the outset, the structure of the CH3 domain is of special interest because it mediates protein-protein interactions that are crucial for regulation of transcription. The interaction with a small peptide (eight residues) from p53, a known activator, with CH3 was investigated with NMR. A specific interaction was found by following shifts in both the backbone and side chains in HSQC spectra. The largest chemical shifts were observed for a small number of residues in three of the helices so that the portion of CH3 interacting with the ligand could be identified. The dissociation constant for the interaction of the peptide and CH3 is about 300 μM.

TABLE 7-1. NMR Restraints and Statistics for the Structure of theTAZ2 (CH3) Domain

A. NMR restraints	
Total NMR restraints	1030
Distance restraints	846
Intraresidue	145
Sequential	286
Medium range	252
Long range	163
Total dihedral angle restraints	184
B. NOE violations	
Average violation (Å)	0.11 ± 0.04
Maximum violation (Å)	0.25
C. Deviations from ideal covalent geometry	
Bond lengths (Å)	0.0057 ± 0.0001
Bond angles (deg)	2.41 ± 0.02

Reference 3

This study indicates the variety of information that can be obtained with NMR: The structure of an important biological molecule was determined; the specific residues binding the Zn^{2+} were determined; and the binding region for the interaction of CH3 with a regulatory molecule was identified.

PROTEIN-DNA INTERACTIONS

Proto-oncogenes are segments of DNA that code for proteins that have a normal function, but can be mutated or altered to become cancer-causing oncogenes. One of these, c-myc, encodes a nuclear protein involved in the regulation of transcription. The expression of c-myc is in turn regulated by FUSE binding protein (**F**ar-**U**p**S**tream **E**lement). This protein binds to single-stranded DNA (ssDNA) that is about 1500 base pairs upstream from the c-myc promoter. This protein, FBP, contains four homologous repeats (KH) that are separated by linkers of varying lengths (4, 5). The minimal ssDNA binding domain is designated as KH3-KH4. The structure of the KH3-KH4 domain bound to ssDNA (20–29 nucleotides long) has been solved using multidimensional NMR (6). The total molecular weight of the complex is about 30,000, and 3,153 NMR restraints were used to determine the structure. A representation of the best-fit structure for the KH3-ssDNA is shown in Figure 7-3 (see color plates). The ssDNA is shown as a ball-and-stick model, whereas the protein chain is shown to delineate the overall fold and the α-helices and β-sheets.

The protein fold has three α-helices packed onto a three-stranded anti-parallel β-sheet. The ssDNA binds in a groove formed by helices 1 and 2. The center of the groove is hydrophobic and the edges are hydrophilic so that the DNA bases point toward the center and the sugar phosphates toward the left-hand side of the protein. As expected, recognition of the DNA involves a number of intermolecular hydrogen bonds. This protein binding scaffold for nucleic acid binding can be fine-tuned for either ssDNA or RNA. The KH3 and KH4 domains do not interact with each other. The flexible linker between the domains is 30 amino acids long. NMR relaxation time measurements indicate that the two domains wobble with respect to each other, with time constants of nanoseconds. The flexibility of the domains appears to be of functional significance. If the flexibility of the motion is restricted by deleting four of the DNA bases in the intervening ssDNA between domains, the *c-myc* expression is reduced by a factor of four whereas adding four bases has no effect.

What is the significance of this structure? The functional concept is that DNA transcription can be controlled at a significant distance along the DNA by recognition of the ssDNA that is formed during transcription. The structure determined shows that the regulatory element (FUSE) upstream from the *c-myc* promoter is recognized specifically and forms a very stable complex with the FBA. This work, when combined with functional studies, can be used to construct a mechanism for transcriptional control. Furthermore, the over-expression of *c-myc* has been linked to cancer: the loss of FBP prevents *c-myc* expression and halts cellular proliferation. The disruption of the FBP-ssDNA interaction, therefore, is a potential target for cancer therapy.

DYNAMICS OF PROTEIN FOLDING

Understanding how proteins fold into their native conformation is of central importance in biology. First, protein folding is a necessary part of cellular metabolism and development. Second, understanding how proteins fold provides information about the intramolecular forces in proteins. Many proteins fold and unfold very fast, in times less than a second, so that study of the dynamics of protein folding requires special methods, including the use of NMR (cf. 7).

A prototypical fast folding protein is the N-terminal domain of bacterio-phage λ repressor, a crucial molecule in gene regulation (cf. 8). The structure of the N-terminal domain has been determined both with crystallography and NMR, and it is essentially identical to the full-length version of λ repressor. The protein unfolds in a simple two state process, both thermally and in the presence of urea. The folding reaction can be written as:

$$N \underset{k_f}{\overset{k_u}{\rightleftarrows}} D \qquad\qquad (7\text{-}1)$$

where N and D are the native and denatured states, and k_f and k_r are the first-order rate constants for folding and unfolding, respectively.

The aromatic region of the NMR spectrum is quite different for the native and denatured states as shown in Figure 7-4, where the NMR spectra are shown at various urea concentrations (9–11). Both the native and denatured states show well-resolved spectra. Some of the central peaks are in fast exchange at all urea concentrations so that only sharp resonances are observed, and the resonance most downfield is in slow exchange. However, some of the peaks show broadening at intermediate urea concentrations because the rate of interconversion of the native and denatured states is comparable to the chemical shift between the native and denatured resonances, about 100 cps.

The resonances chosen for analysis are associated with two specific tyrosine residues. The analysis of the data calculated the line shapes from theoretical considerations and compared the calculated and experimental line shape with varying rates of reaction until the two matched. Basically, this measured the effect of the reaction rate on the spin-spin relaxation time, T_2, and yielded the relaxation time for the reaction in Eq. 7-1, τ (Eq. 6-8). The reciprocal of the relaxation time for the chemical reaction is the sum of the two rate constants (12). The proportion of the protein present in the native and denatured states can be derived from the line shape analysis since it is directly proportional to the strength of the resonance for each state. Thus, the individual rate constants can be directly determined. The rate constants vary linearly with the urea concentration but are in the range of 10^2–$10^3 s^{-1}$. When extrapolated to zero urea concentration, the rate constants for folding and unfolding are $3600 s^{-1}$ and $27 s^{-1}$ at 37°C, respectively, for the particular variant of the λ repressor studied (9).

Why does this protein fold so fast? It is relatively small, only 79 amino acid residues, but some proteins of this size fold much slower. A unique feature of this protein is that the only significant secondary structural elements are α-helices, which are known to wind and unwind in microseconds or less. In contrast, β-sheet structures form and break down considerably slower. The detailed folding mechanism has been explored further by site-specific mutagenesis of amino acid residues. The results obtained suggest that minor modifications, such as changing two glycines to alanines or disruption of a single hydrogen bond, can have significant effects on the folding mechanism. The results further suggest that formation of one of the five helical stretches may be a crucial slow step in the folding mechanism.

Knowledge of the kinetics of fast folding proteins provides unique information about the folding mechanism, and NMR provides one of the few methods available for studying these very fast reactions.

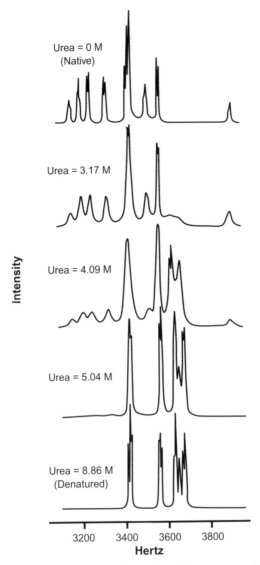

Figure 7-4. NMR spectra of the aromatic region of λ repressor at various concentrations of urea. Differences between the native and denatured (*top* and *bottom*) can be seen. The sharp peaks in the middle are in fast exchange. At intermediate urea concentrations line broadening of several of the peaks can be seen because of chemical exchange. Analysis of the line broadening permitted determination of the rate constants for the interconversion of the native and denatured protein. Reprinted in part with permission from J. K. Myers and T. G. Oas, *Biochemistry* **38**, 6761 (1999). © 1999 by American Chemical Society.

RNA FOLDING

The folding of RNA into functional structures is of obvious importance to biology. The mechanism of folding appears to be more complex than for proteins (cf. 13). A number of structures exist that have similar free energies so that molecules can become trapped in nonfunctional structures. Furthermore, because RNA molecules are highly charged, the structures formed are very dependent on the ionic environment, particularly on the concentration of Mg^{2+}. A frequently invoked mechanism is that two major structural changes occur in the folding reaction. First, stable secondary structures form on the microsecond time scale. Secondary structure is defined as local structural elements such as helices, etc., due to base pairing. The second stage is formation of tertiary folding that brings the secondary structural elements together. This is similar to the mechanism of protein folding discussed above. The study described below suggests that this picture is too simple in many cases.

We have previously discussed ribozymes (Chapter 2). The folding of a self-splicing RNA from *Tetrahymena* has served as a prototypical example of RNA folding (14, 15). This group I intron consists of two large domains, labeled P4-P6 and P1-P2/P3-P9. The stable P4-P6 domain folds independently, and its crystal structure has been determined by x-ray crystallography (16, 17). Within this structure are three helices, labeled p5abc, and the structure of this RNA has been studied with NMR (18). The P5abc RNA can fold independently, and its structure is amenable to determination by NMR.

To arrive at a structure, the one-dimensional proton NMR spectrum was obtained for the imino protons, protons that are involved in the hydrogen bonding of base pairs. An A-U base pair has one imino proton, and a G-C base pair has two imino protons. A two-dimensional NOESY study was then carried out to determine which imino protons were close to each other. Interpretation of these spectra required assignment of each of the resonance peaks in the one-dimensional spectrum to specific bases in the RNA. The imino protons if guanine and uracil give rise to very sharp NMR resonances in the 10–15 ppm range. This suggests that a single conformation is present, and no aggregation is occurring. In the NOESY spectrum, the diagonal peaks corresponded to the one-dimensional spectrum, and each cross-peak connected two diagonal peaks, which implies the two protons are within 5 Å of each other. Neighboring base pairs give rise to NOEs and helped establish the assignments of specific resonance peaks and the secondary structure of the RNA.

The results are summarized in Figure 7-5, where the secondary structure of the 56 nucleotide RNA is shown. The arrows indicate the connectivitiy established by the NOESY experiments, often called a "NOE walk." Included in the structure are disks that indicate the locations of the imino protons, all of which are involved in base pairs.

The NMR results give a quite clear picture of the secondary structure. Surprisingly, when this structure was compared with that determined with x-ray crystallography, the two structures were found to be different. In the crystals,

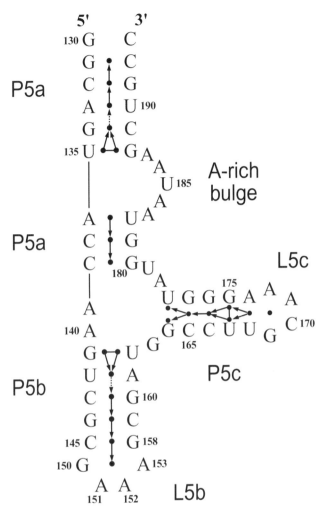

Figure 7-5. Secondary structure of the 56-nucleotide RNA ribozyme fragment. The disk between each base pair represents the imino protons that are observed by NMR. The arrows represent the connectivity established by the NOESY experiments. The dotted arrow NOE was not observed because the resonances could not be resolved. Reproduced with permission from M. Wu and I. Tinoco Jr., *Proc. Natl Acad. Sci. USA* **95**, 11555 (1998). © 1998 National Academy of Sciences, USA.

the hydrogen bonding of six base pairs are broken (two G-C, one A-U, and three G-U) and four new G-C base pairs are formed, as well as two non-standard A-U pairs. In addition, a tetraloop is disrupted, and other changes occur. The calculated free energy of the secondary structure of the RNA crystal structure actually is higher than that of the NMR secondary structure.

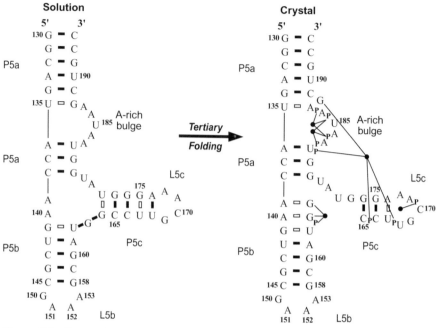

Figure 7-6. Cornparison of the structures of the RNA ribozyme fragment determined by NMR and crystallography. The tertiary folding is caused by the five Mg^{2+} ions shown in the crystal structure as filled circles. The solid bars between bases represent Watson-Crick pairing, and the open bars represent non-Watson-Crick base pairing. Note the changes in base pairing caused by the presence of Mg^{2+}. Reproduced with permission from M. Wu and I. Tinoco Jr., *Proc. Natl Acad. Sci. USA* **95**, 11555 (1998). © 1998 National Academy of Sciences, USA

However, this difference in free energy is more than compensated for by inter-actions within the tertiary structure. These structural differences can be rec-onciled if Mg^{2+} is added to the solution. The addition of this metal results in a conversion to a new structure that is identical to that found in the crystal struc-ture. In fact, two sets of resonances are seen, indicating that the two structures are slowly interconverting. In this case, slow means with a rate constant less than $30\,s^{-1}$. (This is determined by the differences in chemical shifts between the two structures.) In terms of the structure, the interactions of RNA with Mg^{2+} alter both secondary structure and the tertiary structure. This change in structure is shown schematically in Figure 7-6. In the fully folded RNA, a cluster of five Mg^{2+} ions are coordinated to phosphate oxygens, indicating the tertiary folding occurs on a Mg^{2+} core.

The significance of these results in understanding RNA folding is that pre-formed secondary structure cannot be assumed to be the direct precursor of the final tertiary structure (18, 19). In other cases, this has been found to be a valid assumption. It also indicates that many stable conformations are available for RNA within a range of salt and Mg^{2+} concentrations. This makes the delineation of the physiological folding mechanism for RNA more difficult than for proteins where a more restricted number of stable conformations probably exists.

In a follow-up study, site-specific mutations of the P5abc RNA fragment were made to explore the balance between tertiary and secondary structures (20). In particular, point mutations designed to disrupt the secondary structure greatly affected the Mg^{2+}- dependent folding into the new structure. Moreover, two single-point mutations are sufficient to prevent the rearrangement of the secondary structure observed in the crystal structure. However, if the same experiments are done with the P4-P6 domain, formation of the tertiary structure is sufficient to alter the secondary structure of P5abc, as observed with the native P5abc RNA, even with the point mutations. The message of this work is that the balance between tertiary and secondary structure is very delicate in RNA. It depends on both the ionic conditions and the context of the RNA domains. This is likely to be a general feature of RNA structure/ folding.

LACTOSE PERMEASE

Transport proteins are integral membrane proteins that are responsible for the flow of many metabolites across the cell membrane. Two broad classes of transport mechanisms exist, *facilitated diffusion* and *active transport*. Facilitated diffusion depends on a concentration gradient, and the molecules being transported flow from a higher to a lower concentration. Selectivity is due to the size of the pore and/or gating of the channel by a stimulus. For example, Gram-negative bacteria contain several porins, 34–38 kDalton proteins, in their outer membranes. The porins permit molecules and ions with a molecular mass of about 600 to enter. The molecules that are transported include maltodextrins, sugar phosphates, and chelated iron. Glucose and HCO_3^- are also generally transported across membranes by facilitated diffusion. In active transport, material is carried across the membrane against a concentration gradient, from low to high concentration. Active transport must be coupled to an electrochemical gradient created by a free energy favorable process such as the hydrolysis of ATP.

Lactose permease (LacY) is responsible for all of the translocation reactions carried out by the galactoside transport system in *E. coli*. It couples the free energy associated with the collapse of a proton gradient, that is, transport of protons, with the energetically uphill translocation of galactosides against a concentration gradient. The lactose permease from *E. coli* has been extensively studied (cf. 21), and ESR has been one of the tools used to elucidate its structure in the membrane. The protein has been purified, reconstituted into proteoliposomes, and shown to be solely responsible for the galactoside transport.

The crystal structure of lactose permease is now known (22), but prior to the structure determination, the primary features of the molecule had been determined by less direct methods such as mutagenesis, chemical cross-linking, and ESR. Lactose permease has a molecular weight of about 45,500, and the structure contains 12 α-helices passing through the membrane, linked by loops on each side of the membrane. The structure is shown schematically in Figure 7-7. A hydrophilic cavity is formed between the helices that alternately faces the inside and outside of the cell as the sugar is transported. We will not be concerned about the detailed mechanism of transport here, but rather with the structural information that was obtained from a series of ESR studies. In simplistic terms, the mechanism consists of a series of sugar and proton bindings that result in the conformation of the protein switching between conformations that expose the hydrophilic cavity to the appropriate side of the membrane. The structural models for the inward and outward facing conformations are shown schematically in Figure 7-8: note the complex arrangement and distortions of the helical rods that make up the structure.

Site-directed spin labeling of lactose permease can be carried out by introduction of a single cysteine residue into a protein with no cysteines, and then labeling the thiol with a nitroxide spin label. The protein with no cysteines (prepared by site-specific mutagenesis) retains its transport activity. In the initial set of experiments (23), three derivatives were prepared with cysteines at amino acid positions 148 (helix V) and 228 (helix VII), 148 and 226 (helix VII), or 148 and 275 (helix VIII). A nitroxide spin label was then covalently linked to the cysteines. The ESR spectra showed relatively broad lines for all

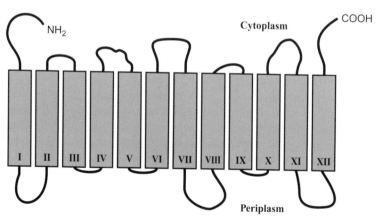

Figure 7-7. Secondary structure model of lac permcase. The permease has a hydrophilic N terminus, followed by 12 α-helical hydrophobic domains that are connected by hydrophilic loops, and a hydrophilic C-terminus. Adapted with permission from J. Wu, J. Voss, W. L. Hubbell, and H. R. Kaback, *Proc. Natl. Acacl. Sci. USA* **93**, 10123 (1996). © 1996 National Academy of Sciences, USA.

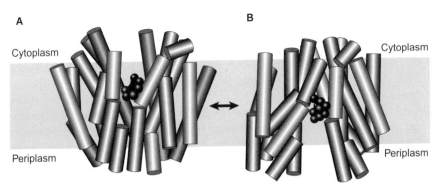

Figure 7-8. Schematic representation of the inward- and outward-facing conformations of lac permease. The mechanism of lactose transport involves alternating these conformations. Lactose is shown as a space-filling molecular structure. Adapted with permission front J. Abramson, I. Smirnova, V. Kasho, G. Verner, H. R. Kaback, and S. Iwata, *Science* **301**, 610 (2003). © 2003 AAAS.

three pairs, suggesting the nitroxides were relatively immobile due to strong interactions with their environment, nearby protein and membrane. Additional broadening could be detected at low temperatures for the 148/228 and 148/275 pairs due to close proximity of the nitroxides (spin-spin interactions). Analysis of this additional broadening indicates the spin labels are within about 15 Å of each other. These results suggest that position 148 in helix V is in close proximity to position 228 in helix VII and position 275 in helix VIII. Chemical cross-linking experiment indicate that position 148 is closer to helix VII than to helix VIII. On the other hand, the lack of spin-spin interactions between positions 148 and 226 suggests that cysteine 226 is on the opposite face of helix VII from cysteine 228.

In a second study (24), a high-affinity Cu^{2+} binding site was created on lactose permease by replacing Arg 301 (helix IX) and Glu 325 (helix X) with His residues. In addition, a series of proteins with single cysteines at various positions in helices II, V, or VII were prepared, and nitroxide spin labels were attached to the cysteines. The ESR spectra of the nitroxides varied, depending on their locations. Those interacting within the protein structure were broad, indicating restricted rotation, whereas those directed away from the interior of the protein were much sharper, indicating freer rotation. When Cu^{2+} was added, some of the lines broadened due to interaction of the unpaired electron of the metal ion with the spin label. This permitted a mapping of the cysteines with respect to the metal ion. When these results were combined with those from chemical cross-linking, fluorescence, and other techniques, a model for the helix packing could be developed. Additional experiments of this type utilized the interaction of Gd(III) with spin labels to provide confirmation that helix V lies close to both helices VII and VIII (25).

Single cysteines also were introduced at positions 126–156, and the cysteines were derivatized with nitroxide spin labels (26). The dynamics of these labels varied from highly mobile to highly immobilized, as judged by the broadness of the spectra. These results can be interpreted in terms of structure in that enhanced mobility can be interpreted as increased accessibility of the side chain. In addition, spectral broadening due to addition of paramagnetic species (potassium chromium oxalate and oxygen) provided further information about the accessibility of the spin labels to the solvent. These experiments provide information about both the structure and the dynamics of the protein side chains.

These extensive ESR studies provide detailed structural information about a membrane-bound protein. Such information is difficult to obtain because membrane-bound proteins are not readily accessible to the large cadre of techniques that can be applied to soluble proteins. In this case, a crystal structure has been obtained that is in reasonable accord with the more indirect methods used. The studies of lactose permease serve as a useful paradigm for investigation of the structure of membrane-bound proteins.

CONCLUSION

These selected examples of the application of magnetic resonance methods to biology are a small fraction of the many interesting investigations that have been carried out. Note that in each case, important questions were being asked about structure and function. In point of fact, the biological problem is the most significant aspect of these studies, and magnetic resonance is one of many spectroscopic/biochemical tools that can be used for the elucidation of these problems.

REFERENCES

1. A. Giordano and M. L. Avantaggiati, *J. Cell Phys.* **181**, 218 (1999).
2. R. H. Goodman and S. M. Smolik, *Genes Dev.* **14**, 1553 (2000).
3. R. N. De Guzman, H. Y. Liu, M. Martinez-Yamout, H. J. Dyson, and P. E. Wright, *J. Mol. Biol.* **303**, 243 (2000).
4. R. Duncan, L. Bazar, G. Michelotti, T. Tomonaga, H. Krutzsch, M. Avigan, and D. Levens, *Genes Dev.* **8**, 465 (1994).
5. C. G. Burd and G. Dreyfuss, *Science* **265**, 615 (1994).
6. D. T. Braddock, J. M. Louis, J. L. Baber, D. Levens, and G. M. Clore, *Nature* **415**, 1051 (2002).
7. J. K. Myers and T. G. Oas, *Annu. Rev. Biochem.* **71**, 783 (2002).
8. R. T. Sauer, S. R. Jordan, and C. O. Pabo, *J. Mol. Biol.* **227**, 177 (1990).
9. G. S. Huang and T. G. Oas, *Proc. Natl. Acad. Sci. USA* **92**, 6878 (1995).

10. R. E. Burton, G. S. Huang, M. A. Daugherty, P. W. Fullbright, and T. G. Oas, *J. Mol. Biol.* **263**, 311 (1996).

11. J. K. Myers and T. G. Oas, *Biochemistry* **38**, 6761 (1999).

12. G. G. Hammes, *Thermodynamics and Kinetics for the Biological Sciences*, John Wiley and Sons, New York, 2000, p. 88–89.

13. P. Brion and E. Westhof, *Annu. Rev. Biophys. Biomolec. Struct.* **26**, 113 (1997).

14. P. P. Zarrinker and J. R. Williamson, *Nat. Struct. Biol.* **3**, 432 (1996).

15. B. Scalvi, M. Sullivan, M. R. Chance, M. Brenowitz, and S. A. Woodson, *Science* **279**, 1940 (1998).

16. J. H. Cate, A. R. Gooding, E. Podell, K. Zhou, B. L. Golden, C. E. Kundrot, T. R. Cech, and J. A. Doudna, *Science* **273**, 1678 (1996).

17. J. H. Cate, A. R. Gooding, E. Podell, K. Zhou, B. L. Golden, A. A. Szewczak, T. R. Cech, and J. A. Doudna, *Science* **273**, 1696 (1996).

18. M. Wu and I. Tinoco, Jr., *Proc. Natl. Acad. Sci. USA* **95**, 11555 (1998).

19. D. Thirumalai, *Proc. Natl. Acad. Sci. USA* **95**, 11506 (1998).

20. S. K. Silverman, M. Zheng, M. Wu, I. Tinoco Jr., and T. R. Cech, *RNA* **5**, 1665 (1999).

21. J. Abramson, I. Smirnova, V. Kasho, G. Verner, S. Iwata, and H. R. Kaback, *FEBS Lett.* **555**, 96 (2003).

22. J. Abramson, I. Smirnova, V. Kasho, G. Verner, H. R. Kaback, and S. Iwata, *Science* **301**, 610 (2003).

23. J. Wu, J. Voss, W. L. Hubbell, and H. R. Kaback, *Proc. Natl. Acad. Sci. USA* **93**, 10123 (1996).

24. J. Voss, W. L. Hubbell, and H. R. Kaback, *Biochemistry* **37**, 211 (1998).

25. J. Voss, J. Wu, W. L. Hubbell, V. Jacques, C. F. Meares, and H. R. Kaback, *Biochemistry* **40**, 3184 (2001).

26. M. Zhao, K.-C. Zen, J. Hernandez-Borrell, C. Altenbach, W. L. Hubbell, and H. R. Kaback, *Biochemistry* **38**, 15970 (1999).

CHAPTER 8

MASS SPECTROMETRY

INTRODUCTION

The field of mass spectrometry does not fall within the rubric of spectroscopy as it does not involve the interaction of matter with radiation. However, its use in biology has become sufficiently prevalent that the basic concepts and their applications merit consideration in this text. The basic variable parameter in spectroscopy is wavelength whereas in mass spectrometry it is m/z, the ratio of the mass of a particle to its charge. In very simple terms, a mass spectrometry experiment can be divided into three main steps: ionization, mass analysis, and ion detection. Although modern mass spectrometry experiments are more complicated than suggested by this simple analysis, these three steps are always involved. Importantly, the only environment in which ions are stable for a sufficient time to analyze readily is a vacuum, and a good vacuum is an essential part of a mass spectrometer. This chapter constitutes an introduction to the field. More comprehensive treatments can be found in references 1–3.

MASS ANALYSIS

The basic principles underlying mass spectrometry can be understood by considering the motion of a charged particle in electric and magnetic fields.

Spectroscopy for the Biological Sciences, by Gordon G. Hammes
Copyright © 2005 John Wiley & Sons, Inc.

The kinetic energy of a charged particle in an electric field is given by the relationship.

$$\frac{1}{2}mv^2 = zV \tag{8-1}$$

where m is the mass, v is the velocity, z is the charge, and V is the applied voltage. The trajectory of an ion in a magnetic field, H, is an arc of radius r, as shown in Figure 8-1. The conservation of angular momentum requires that

$$mv^2/r = zvH \tag{8-2}$$

Eliminating the velocity between these two equations gives

$$z/m = H^2r^2/(2V) \tag{8-3}$$

This relationship shows that in constant magnetic and electric fields, the trajectory will be the same for all ions with the same ratio m/z. Thus, if the magnetic and electric field are kept constant, the mass spectrum can be collected by determining the positions of the ions after they have passed through the magnetic field. Alternatively, the magnetic field can be scanned to determine the mass spectrum.

The ultimate goal of the mass analysis is to determine m/z with the smallest possible error and to analyze a broad range of m/z. Unfortunately, these goals are often in conflict so that several different methods of mass analysis are used. The resolution of a mass spectrometer is defined as m/Δm. For example, a mass resolution of 1000 can distinguish between ions with m/z of 1000 and 1001.

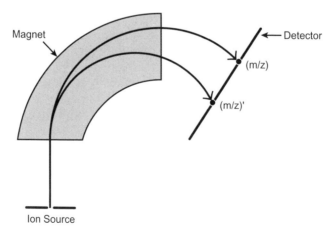

Figure 8-1. Schematic representation of the trajectory of ions with different m/z values in a magnetic field. In this illustration, m/z > (m/z)'.

The simplest type of mass spectrometer utilizes only a magnetic analyzer. As is evident from Eq. 8-2, the path of an ion in a magnetic field is different for every value of m/z. If Eq. 8-2 is solved for r, we find that

$$r = (1/H)(2mV/z)^{1/2} \tag{8-4}$$

The path for two ions with different m/z passing through a magnetic field is included in Figure 8-1. The resolution of the separation can be determined by taking the natural logarithm of Eq. 8-4 and differentiating:

$$\Delta r/r = (1/2)(\Delta m/m) + (1/2)(\Delta V/V) \tag{8-5}$$

Thus, in order to sharply focus the ions, thereby decreasing the error in m, it is important for all ions to be homogeneous in energy, that is, the last term must be small. This can be partially accomplished by using a high voltage (V), typically 8000 volts, but usually the magnetic focusing is coupled with an electrostatic analyzer in order to produce ions that are homogeneous in energy (small ΔV). This can be done by subjecting the ions to a constant voltage either prior or subsequent to the magnetic field. Instruments of this type, double-focusing magnetic sector mass spectrometers, have high resolution and a mass range up to about 15,000 daltons. They are also very expensive.

The quadrupole mass spectrometer is the most widely used, primarily because of its relatively low cost. The separation of ions is accomplished by utilizing electric fields only. The mass analyzer consists of four metal rods, as shown in Figure 8-2. The trajectory of the ions is between the four rods. The rods are electrically connected in pairs, A to A and B to B in the figure. A constant voltage of opposite sign is applied to the A and B rods. An oscillating voltage is superimposed on the constant voltage, with the phase differing by 180° between A and B. The quadrupole serves as a mass filter: only ions with a specified value of m/z can get through the filter. Other ions collide with the rods and do not reach the detector. To obtain a mass spectrum, the applied electric fields are varied, thus allowing ions with different values of m/z to be detected. The resolution of quadrupole instruments is not as good as double-focusing magnetic sector instruments, 10,000 versus 100,000, but it is significantly less expensive and can tolerate relatively high pressures. The latter feature is important if the ions are generated by electrospray, a frequently used technique that will be discussed later.

The quadrupole mass spectrometer has been coupled with an "ion trap." With these instruments, the ions are held within the quadrupole and manipulated before proceeding to the detector. By holding the ions in the trap, both a time and space dimension are available. This improves the resolution and sensitivity. In simple terms, the ions are physically trapped between electrodes and subjected to both constant and oscillating electric fields such that ions of specific m/z precess within the trap. The magnitudes of the fields are increased, thereby causing ions of specific m/z to be ejected from the trap.

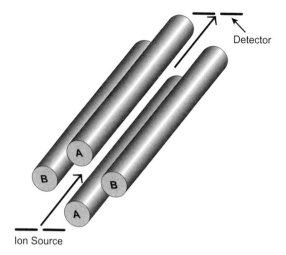

Detector

B

A

B

A

Ion Source

Figure 8-2. Schematic representation of a quadropole mass analyzer. The ions pass through the middle of four parallel rods. The A rods are connected and have the same DC and superimposed radiofrequency voltages. The B rods are also connected and have a DC voltage opposite in sign to the A rods and the radiofrequency voltage phase shifted 180° relative to the A rods.

The time-of-flight mass spectrometer selects ions by measuring the time of arrival of the ion at the detector. Eq. 8-1 can be rearranged to give.

$$v = (2Vz/m)^{1/2} \tag{8-6}$$

This equation predicts that the lighter the ion, the faster it will arrive at the detector. In order for this method to work, the ions must enter the flight tube at the same time. This is accomplished by generating ions in short bursts. The difference in the time of arrival of ions is not great, typically in the microsecond range. Thus, a complete spectrum can be determined in very short times. In some cases, reflectors are used to improve the sensitivity and resolution. This is done by slowing the ions with a series of electric field "lenses" until they essentially stop and then accelerating them in the opposite direction. The reflection increases the path length traveled thus providing better separation of the time-of-flight, and the lenses focus the ions with a specific m/z by reducing the spread in kinetic energies for a given ion. The time-of–flight instrument has the advantage of essentially unlimited mass range and high sensitivity.

Other methods exist for mass analysis but are not discussed in detail here. Most notably, Fourier transform methods using ion cyclotron resonance have been developed that permit very high precision mass determinations. The ions are inserted into a small volume in the cyclotron, and a large magnetic field is applied so that the ions precess in circular orbits according to Eq. 8-2. The ions

TABLE 8-1. Characteristics of Mass Analyzers

Method	Mass Range (Dalton)	Resolution
Magnetic Sector	15,000	200,000
Quadrupole	4,000	4,000
Quadrupole ion trap	100,000	30,000
Time-of-flight	Unlimited	15,000
Fourier transform-ion cyclotron resonance	$>10^6$	$>10^6$

Adapted from reference 1.

are constrained to the cell by an electric field applied to front and rear plates of the sample cell. If a pulsed electric field is applied at a frequency matching the precession frequency of the ions, energy is absorbed, analogous to a magnetic resonance experiment. The ions then transmit a radio frequency current at the detector plates that contains the frequency components of each of the ions. This is converted to a free ion decay signal (analogous to the free induction decay in NMR) that can be Fourier transformed to the mass spectrum of all of the ions, thus permitting detection over a wide mass range.

A summary of the various mass analyzers, along with their approximate range of m/z and resolution is given in Table 8-1.

TANDEM MASS SPECTROMETRY (MS/MS)

Tandem mass spectrometry couples two (or more) mass analyzers to obtain additional information about the sample in question. Three steps are typically involved in MS/MS analysis: mass selection, fragmentation, and mass analysis. The first step is the selection of a specific ion for further study. In a second step, the selected ion then undergoes fragmentation, usually through collisions with neutral gas atoms. The ion fragments are then analyzed by a second mass analyzer. Because modern ionization methods produce very little fragmentation, the ion selected is often similar to the ion of the parent compound. Selection of a specific ion and determination of its fragmentation pattern may be important for its molecular identification.

A variety of different MS/MS instruments are available. For example, a triple quadrupole instrument uses the first quadrupole for ion selection, the second for collision induced dissociation, and the third for analysis of the fragments produced. Ion traps can also be incorporated at a relatively low cost. Similarly multiple sector instruments are available that use a series of magnetic and electric fields for analysis. As might be expected, multiple sector instruments are quite expensive. Time-of-flight and Fourier transform-cyclotron resonance instruments also are available. The use of MS/MS is particularly useful for biological systems to identify and characterize substances uniquely.

ION DETECTORS

The two most common methods of detecting ions after they have been sorted by the mass spectrometer are electron multipliers and photomultipliers. With electron multipliers, the ion strikes a dynode that emits secondary electrons. Typical dynode surfaces are BeO, GaP, and CsSb. These secondary electrons are accelerated by a voltage and attracted to a second dynode that emits more electrons. This process continues through a series of dynodes, resulting in a cascade of electrons. The resulting current can be read with standard technology. Electron multipliers are very sensitive: a signal amplification factor of 10^6 can be readily obtained.

Photomultiplier detectors operate in a similar fashion except that the ion first strikes a phosphorous screen. The phosphorous screen releases photons, which are detected by a photomultiplier. The photomultiplier also has a series of dynodes and causes a cascade of electrons when the light strikes it. This type of detection is commonly called scintillation counting and is often used to measure radioactivity quantitatively. The amplification factor is similar to electron multipliers. A major advantage of photomultipliers is that they have significantly longer lifetimes than electron multipliers. The lifetime of electron multipliers is limited by contamination/damage of the surface that the ions strike. Nonetheless, electron multipliers are currently the most common devices used.

Detection of the ion signal is often done by a point detector so that only a single type of ion is detected, that is, a single m/z. However, array detectors also are available. Array detectors consist of a linear arrangement of detectors so that multiple ions can be detected simultaneously.

IONIZATION OF THE SAMPLE

The first step in the analysis is to ionize the sample into the vacuum of the mass spectrometer. A variety of methods are used to ionize the sample, and we will only deal with a few of them. Currently, the two most widely used methods for macromolecules are MALDI, **M**atrix **A**ssisted **L**aser **D**esorption/**I**onization, and ESI, **E**lectro**S**pray **I**onization, but a few other methods that are commonly used for relatively small molecules also will be discussed.

Electron ionization vaporizes the sample into the mass spectrometer and then bombards the vapor with a beam of high energy electrons, 50–100 eV. This is typically accomplished by thermal evaporation from a probe containing the sample. The high-energy electrons are produced from a filament and acceleration of the electrons through a large electric field. In order to produce ions, the electrons must have an energy greater than the ionization energy of the molecule, M, being studied. This process can be written as:

$$M + e^- \rightarrow M^{+\cdot} + 2e^- \tag{8-7}$$

The positive ions formed, M^+, are referred to as odd-electron molecular ions or radical cations. Because the ionization energy for most molecules is only about 5 eV, the molecular ion produced usually has an excess of energy and fragments. The fragmentation pattern provides a "fingerprint" for each molecule. Libraries of fragmentation patterns are available for the identification of unknown compounds. The primary drawback of this method is that relatively stable and volatile compounds are required. In practice this limits the molecular weight of samples to less than about 1000 daltons.

The technique of fast ion bombardment (FAB) was developed in the 1980s and permits substances with molecular weights of 4000 or greater to be analyzed routinely. The principle of the method is to place a sample that is dissolved into a matrix on the tip of a metal probe. The sample is then bombarded with a stream of fast (high temperature) Ar or Xe atoms, and molecular ions are produced, predominately by the loss or gain of an H atom. (In some cases, a beam of Cs^+ is used.) The key to this method is the matrix, which is a viscous liquid and relatively inert. Typical examples are *m*-nitrobenzyl alcohol and glycerol. The sample is dissolved in a solvent that is miscible with the matrix. The matrix absorbs most of the high-energy atoms and produces a high-temperature, high-density gas within a small volume. Various matrix ions are produced, protonated, and deprotonated, and these react with the sample molecule, M, to produce ions:

$$M + matrix - H^+ \rightarrow MH^+ + matrix$$
$$MH + matrix^- \rightarrow M^- + matrix - H \qquad (8\text{-}8)$$

Thus, both negative and positive sample ions are produced. The energy of the ions is considerably less than produced by electron ionization, but some fragmentation occurs. A schematic representation of FAB is shown in Figure 8-3.

The concept behind MALDI is similar to FAB (Fig. 8-3). The sample is embedded in a matrix, and an external source is used to convert the sample to ions. In the case of MALDI, the sample is embedded into a solid crystalline compound, and a laser, operating at a wavelength at which the matrix strongly absorbs, is used to desorb and ionize the sample. A laser with a high-energy output is needed: both uv and ir lasers have been used. The matrix is essentially a "solvent" for the sample, and the sample forms a microcrystal with the matrix. In practice, the sample is dissolved into a small amount of solvent (water or water plus organic solvents, typically), and the sample is then mixed with a concentrated solution of the matrix. Just a picomole of sample is required and the ratio of sample to matrix is typically 1/1000. A variety of materials has been used for matrices, for example, cinnamic acid, succinic acid, and urea. The mixture is then evaporated and introduced into the mass spectrometer. Short laser pulses, typically 10–20 ns duration, with about 10^6 W/cm^2 power are used. The exact mechanism for producing ions is not well understood, but the general idea is that the matrix is ionized, and the ions produced

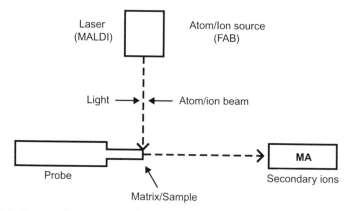

Figure 8-3. Schematic representation of FAB (*right*) and MALDI (*left*) methods of ionization. In both cases, the sample is embedded in a matrix. In the case of FAB, the ions are created by bombardment with an atom or ion beam, and in the case of MALDI by a pulsed laser. The secondary ions are then injected into the mass analyzer.

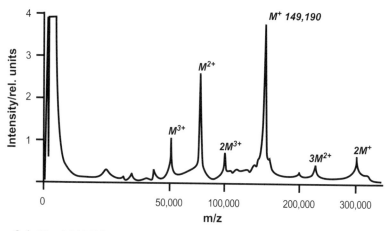

Figure 8-4. The MALDI mass spectrum of a monoclonal antibody. Reprinted from F. Hillenkamp and M. Karas, Mass Spectrometry of Peptides and Proteins by Matrix Assisted Ultraviolet Laser Desorption/Ionization, *Meth. Enzymol.* **193**, 280 (1990). ©1990, with permission from Eisevier.

ionize the sample molecules through a series of proton transfer reactions involving the matrix molecules and the anylate. As with FAB, both positive and negative ions are produced. This method can be used to study macromolecules with molecular weights in excess of 300,000 daltons. Moreover, the sensitivity is excellent, in the pico- to femtomole range, and little fragmentation occurs. A typical mass spectrum of a protein obtained with MALDI is shown in Figure 8-4.

Finally, we consider *electrospray ionization (ESI)*. With this method, a fine spray of highly charged droplets is created by spraying the sample onto the tip of a metal nozzle at approximately 4000 volts. This is done at atmospheric pressure. Solvent is removed through a series of Coulombic explosions. The droplets can also be heated to facilitate evaporation. As a droplet becomes smaller, the electric field density increases, until finally mutual repulsion between the ions causes the ions to leave the droplet. In essence, the electrostatic forces become greater than the surface tension. The ions are then directed into the mass analyzer with an electrostatic lens. An important and unique characteristic of this method is that highly charged ions are produced so that a wide range of m/z values are observed for a single analyte. For example, the mass spectrum of myoglobin obtained with ESI is shown in Figure 8-5. If two adjacent peaks are assumed to differ by only one charge and one proton, the molecular weight can be calculated. This method can be used for molecules of about 100,000 daltons or less and has good sensitivity, in the pico- to femtomole range.

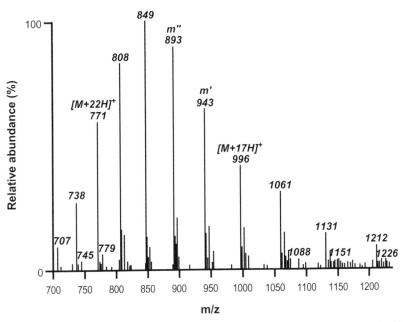

Figure 8-5. An electrospray ionization (ESI) mass spectrum of horse myoglobin. Each peak is characteristic of myoglobin with a different charge and number of protons. From C. Dass, *Principles and Practice of Biological Mass Spectrometry*, Wiley & Sons, New York, 2001, p. 43. Reprinted with permission of John Wiley & Sons, Inc. © 2001.

TABLE 8-2. Ionization Techniques for Biomolecules[a]

Compound	Ionization Method	Ionization Mechanism
Peptides	FAB, MALDI, and ESI	Protonation, deprotonation
Proteins	MALDI and ESI	Protonation
Carbohydrates	FAB, MALDI, and ESI	Protonation, deprotonation, cationization[b]
Oligonucleotides	MALDI and ESI	Protonation, deprotonation, cationization[b]
Small biomolecules	FAB, MALDI, and ESI	Protonation, deprotonation, cationization[b], electron ejection

[a] Adapted from reference 2.
[b] Addition of a cation other than H^+.

SAMPLE PREPARATION/ANALYSIS

Before discussing applications of mass spectrometry, some general comments should be made. Table 8.2 summarizes the most frequently used methodology for biomolecules. The amount of sample required depends on the specific methods used, but generally 5–50 µL of 10–100 µM solutions are adequate. Frequently, a mass spectrometer is coupled to purification procedures such as gas chromatography, high-performance liquid chromatography, or capillary electrophoresis. This provides a high throughput of samples and rapid analysis of unknown mixtures. In some cases, an in-depth study of a single substance is carried out. In this case, the purity is important and extraneous background material should be eliminated, notably salts, if possible.

In order to obtain accurate mass measurements with mass spectrometry, the instrument must be calibrated with standard compounds. This is sometimes done by running a series of calibration curves, but the best procedure is to include an internal mass standard with the sample. This will ensure that fluctuations in instrument response are adequately taken into account. Mass measurements can be made to within 1–500 parts per million. Thus, for example, the accuracy can be within a few tenths of a dalton for a macromolecule of molecular weight of 100,000.

PROTEINS AND PEPTIDES

With the advent of soft ionization methods, the determination of the molecular weight of proteins with mass spectrometry has become routine and has become the method of choice because of its accuracy and the relatively small amount of material required (fmols). The method works best if a pure protein is obtained by conventional methods such as chromatography. With a pure protein, mass spectrometry, usually with MALDI or ESI, can detect differences in a single amino acid quite readily. Thus, for example, mass spectrometry is a confirmatory tool for site-specific mutagenesis. Also if post-translational modifications occur, such as phosphorylation or carbohydrate addition, mass spectrometry can identify and in some cases quantify the modification.

Proteomics is a rapidly developing field that systematically characterizes gene products. Proteomics is concerned with determining the identities of proteins, both known and unknown, and their functions. Tissues or fluids are subjected to protein purification protocols, usually by simple chromatography or gel electrophoresis, and the molecular masses and amino acid sequences of the proteins are then determined by mass spectrometry. This represents the first step in the characterization of gene products. Molecular weight by itself is rarely sufficient to identify a protein so that once proteins have been identified and their molecular weight determined, further studies are carried out. The next step is to take a specific protein, which could be a spot taken from two-dimensional gel electrophoresis of the crude starting material, and subject it to proteolysis with an enzyme such as trypsin. Before the use of mass spectrometry, the resultant peptides had to be separated, often by laborious procedures. However, with mass spectrometry, the trypsin digest can be analyzed directly and the molecular weights of the peptides determined. Since proteins usually have a unique digest, this procedure is often sufficient to determine what the starting protein is. Data banks of trypsin digests for hundreds of proteins and sophisticated software are available to help in the identification process. An example of the mass spectrum of a mixture of peptides is shown in Figure 8-6.

Although molecular weight characterization of peptide fragments after proteolysis is very useful for identifying proteins, the surest identification is to determine the amino acid sequence of the peptides. The amino acid sequence of the protein can then be compared with the vast database of known proteins. In addition to identifying specific proteins, the databases can be searched for homologous proteins, that is, proteins with similar but not identical amino acid sequences.

Determining the sequence of a peptide with mass spectrometry is not as easy as making molecular weight measurements. Basically, the principle is to use fragmentation of the parent peptide as a unique identifier of the sequence. The most frequently used technology is tandem mass spectrometry. With this method, a single peptide can be selected by the initial ion separation, and this peptide can then be subjected to fragmentation and analysis by the second mass analyzer. The amino acid sequence of peptides with molecular weights of up to about 3000 can be readily determined with this technique. If the peptide is large, the amount of useful sequence information that can be generated in the mass spectrometer is limited because of the complexity and large number of fragments produced.

An effective strategy for amino acid sequencing is the use of peptide ladders. With this methodology, a peptide is treated with an exopeptidase (car-

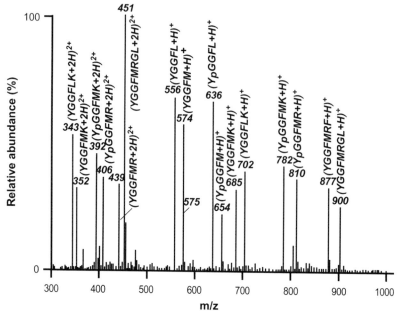

Figure 8-6. The positive ion ESI mass spectrum of a mixture of ten peptides. The masses and amino acid sequences of the major peaks are indicated. Reproduced with permission from C. Dass in B. S. Larsen and C. N. McEwen, eds. *Mass Spectrometry of Biological Materials*, Marcel Dekker, New York (1998), pp. 247–280.

boxypeptidase or aminopeptidase). These enzymes take off one amino acid at a time from the C- or N- terminus. The products of these digestions are then analyzed with mass spectrometry, and the reduction in molecular weight can be used to deduce the amino acid sequence. An example of ladder sequencing is shown in Figure 8-7, where a peptide from HIV protease was sequenced by degradation from the amino terminus. The letter by each mass spectrum peak indicates the amino acid that was found on the N-terminus as the peptide decreased in length (4).

The above sequence of events has not taken into account the fact that proteins often contain disulfide linkages. If this is the case, the protein must first be subjected to a reduction process that converts the disulfides to cysteines, and the cysteines are then typically alkylated so that the disulfides will not reform. Because the molecular weights of the reduced/alkylated and oxidized enzymes are different, mass spectrometry can be used to determine exactly how many disulfide linkages are present.

As the fields of genomics and proteomics expand, mass spectrometry undoubtedly will be a central analytical technique for the characterization of gene products. Both the methodology and database searching techniques are rapidly improving.

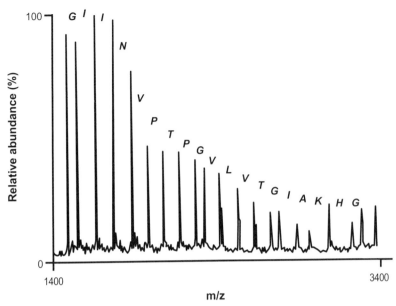

Figure 8-7. Mass spectrum of a mixture of peptides from an HIV protease following sequential removal of an amino acid from the N-terminus of the peptide. The letters indicate the amino acid that was removed from N-terminus. Adapted with permission from B. T. Chait, R. Wang, R. C. Beavis, S. B. H. Kent, *Science* **262** 89 (1993). © 1993 AAAS.

PROTEIN FOLDING

We have previously discussed the transformation of proteins between native and denatured states (Chapters 4 and 7). Mass spectrometry provides a unique tool in such studies. Proteins have many hydrogens that can exchange with water hydrogens. This is a dynamic process and takes place continuously in aqueous solution. The amide hydrogens, that is, those associated with peptide bonds, are of particular interest because their rate of exchange with solvent is dependent on structure and generally occurs over a time range that is readily accessible experimentally. All of the amide hydrogens are replaced by deuterium at approximately the same rate when a denatured protein is put into D_2O. However, for native proteins the rate depends on the environment of each amide hydrogen. For example, if the amide hydrogen is involved in a stable α-helix, the rate of exchange will be slower than if it is freely exposed to solvent. Amide hydrogens directly exposed to solvent exchange rapidly with the solvent. The remaining amide hydrogens can be divided roughly into two classes. One class exchanges due to the making and breaking of local structures: this is often called breathing motions of the protein. The second, slower, exchanging class of hydrogens is associated with the global structure of the protein. They usually do not exchange unless the native structure is denatured. Within each of these classes, subgroups can be found, especially the class associated with local structures (cf. 5 and 6 for reviews of this subject).

For each subgroup the exchange process can be described by the mechanism:

$$NH_{cl} \underset{k_{cl}}{\overset{k_{op}}{\rightleftharpoons}} NH_{op} \xrightarrow{k_{int}} ND \tag{8-9}$$

Here, NH_{cl} and NH_{op} are the closed and open forms of the protonated enzyme: closed means hydrogen exchange cannot occur, and open means hydrogen exchange can occur. ND is the deuterated amide(s), and the rate constants are for the opening, closing, and exchange of the unprotected amide reactions. If the intermediate state is assumed to be in a steady state, the rate constant for the overall exchange reaction is

$$k_{ex} = \frac{k_{op} k_{int}}{k_{cl} + k_{int}} \tag{8-10}$$

Two limiting cases exist. If the rate constant for closure is much faster than the rate constant for the replacement of hydrogen by deuterium ($k_{cl} \gg k_{int}$), $k_{ex} = k_{op} k_{int}/k_{cl}$. If $k_{int} \gg k_{cl}$, then $k_{ex} = k_{op}$. These are called the EX2 and EX1 limits. In the former case, since k_{ex} can be estimated from studies with model compounds, the equilibrium constant k_{op}/k_{cl} can be calculated and is a direct measure of the thermodynamic stability of the structural element under consideration.

If deuterium is substituted for hydrogen, the molecular weight of the protein increases so that mass spectrometry is an excellent tool for studying hydrogen/deuterium exchange. NMR has also been used extensively because it can monitor specific individual amide protons in a single experiment. However, it can not easily detect multiple populations of the same conformation. The combination of NMR and mass spectrometry provides a powerful experimental approach to the study of protein folding, but only a few selected examples studied with mass spectrometry are considered here.

In the EX2 limit, the open and closed forms of the protein are in equilibrium so that, as deuterium is substituted for hydrogen, only a single protein species of increasing molecular weight is observed during the time course of exchange. The starting protein species would have all amide hydrogens and the final protein species all deuteriums, with intermediate species having a specific number of hydrogens and deuteriums. This is shown schematically in Figure 8-8. In the EX1 limit, the opening reaction is rate determining, with rapid exchange following. Therefore, two species should be present during the time course of the exchange, one completely deuterated and the other completely hydrogenated at the amide position. This situation is also shown schematically in Figure 8-8. The shape of the mass spectrum peaks in the EX1

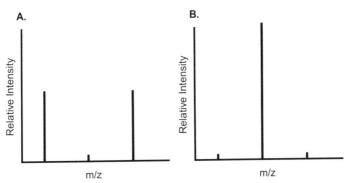

Figure 8-8. Schematic representation of mass spectra for the EXI (A) and EX2 (B) exchange limits midway through the hydrogen/deuterium exchange reaction. The mass of the protein or peptide increases as the amide hydrogen is replaced by deuterium. In the EX1 limit, each protein or peptide molecule has either all hydrogens or all deuterons, whereas in the EX2 limit, each protein or peptide molecule contains both hydrogens and deuterons.

limit can provide information about correlated exchange reactions and other reaction dynamics, but we will not discuss this more complex situation (6).

An extensive study of the incorporation of deuterium into rabbit muscle aldolase has been carried out (7). The protein was incubated in D_2O for a specified length of time. The exchange reaction was quenched by lowering the pD to 2.5 and the temperature to 0°C. Under these conditions the rate of the exchange reaction (k_{ex}) becomes very slow. Denaturant was used to increase the initial rate of exchange—it promotes the formation of open forms. After quenching, the protein was proteolyzed with pepsin, and the peptides were subjected to mass spectral analysis by ESI. The time dependence of the exchange reaction was determined, and it was found that two different peptides incorporated deuterium at different rates. Both protonated and deuterated peptides were found, but intermediate states with both protons and deuterons were not found. This indicated that the EX1 limit is operative. Thus, the mass spectra showed that the EX1 mechanism of exchange was occurring and that the hydrogen/deuterium exchange for the two peptides took place at different rates, indicating differences in kinetic stability within the protein.

Mass spectrometry can also be used to study the exchange reaction in the EX2 limit which provides direct information about the thermodynamic stability of the protein. Mass spectrometry coupled with hydrogen exchange provides a means of rapidly scanning a library of proteins for relative stability (8). The stability of overexpressed proteins in crude cell lysates also can be assessed (9). In the latter case, the increase in mass of the overexpressed protein after exposure to D_2O for a fixed time was determined at different guanidine chloride concentrations. The reaction was quenched by addition to a MALDI matrix. The increase in mass for a number of variants of the protein λ repressor followed the typical sigmoidal behavior for a two-state denaturation, and the relative stability of seven mutants was determined. The results correlated well with those from circular dichroism measurements. The same method can be used to estimate binding constants since ligand binding generally stablizes the native protein configuration. In addition, binding of a ligand or macromolecule may selectively alter the exchange rate for specific regions of the target protein, thus providing information about the region of the protein involved in binding. The ability to rapidly screen many samples is a unique advantage of mass spectrometry for such studies.

OTHER BIOMOLECULES

Although the mass spectrometry literature has been dominated by the study of proteins in recent years, largely because of the genome project and proteomics, many other biomolecules have been studied. Post-translational modification of proteins is a good example. After proteins are synthesized in the cell, they are often modified, for example, by glycosylation and phosphorylation.

The attachment of a carbohydrate to a protein is often important for its function and localization. Carbohydrates also play a significant role in the regulation of physiological processes. Determination of the structure of glycoproteins is challenging, and mass spectrometry can play an important role. If a protein is not homogeneous with respect to glycosylation, multiple molecular weights will be observed. MALDI is especially useful since it generally produces a single charged species and permits the multiple species to be easily sorted. ESI produces multiple ions for each species but has better mass resolution. The molecular mass of the attached carbohydrate can be determined by treating the protein with glycosidases to free the protein of the carbohydrate. If the extent of glycosylation is not great, the nature of the attached group can sometimes be inferred directly. If similar experiments are carried out in conjunction with peptide mapping (proteolysis and separation of peptides), the attachment sites of the carbohydrate can be ascertained. Ultimately, sequencing of the carbohydrate is required for complete characterization, and MS/MS methods are useful in this regard. The coupling of database searching and mass spectrometry is a promising approach for the characterization of carbohydrates. Although the characterization of carbohydrate structures is difficult and not yet well developed, its importance in biology is well recognized (1).

Finally, mention should be made of the application of mass spectrometry to characterizing oligonucleotides and lipids. In the case of lipids, it is a primary tool for structural characterization of the many diverse types of lipids that occur in nature. For nucleic acids, it is especially useful for determining modification of bases. We also have not dwelt on the characterization of metabolites; mass spectrometry is a tool of choice for small molecules because of the small amount of sample required to obtain a complete structure.

REFERENCES

1. C. Dass, *Principles and Practice of Biological Mass Spectrometry*, Wiley & Sons, New York, 2001.
2. G. Siuzdak, *Mass Spectrometry for Biotechnology*, Academic Press, San Diego, CA, 1996.
3. G. Siuzdak, *The Expanding Role of Mass Spectrometry in Biotechnology*, MCC Press, San Diego, CA, 2003.
4. B. T. Chait, R. Wang, R. C. Beavis, and S. B. H. Kent, *Science* **262**, 89 (1993).
5. C. K. Woodward, *Curr. Opin. Struct. Biol.* **4**, 112 (1994).
6. D. M. Ferraro, N. D. Lazo, and A. D. Robertson, *Biochemistry* **43**, 587 (2004).
7. Y. Deng, Z. Zhang, and D. L. Smith, *Amer. Soc. Mass. Spect.* **10**, 675 (1999).
8. D. M. Rosenbaum, S. Roy, and M. H. Hecht, *J. Am. Chem. Soc.* **121**, 9509 (1999).
9. S. Ghaemmaghami, M. C. Fitzgerald, and T. G. Oas, *Proc. Natl. Acad. Sci. USA* **97**, 8296 (2000).

PROBLEMS

8.1. The mass spectrum of the enzyme lysozyme was determined using ESI, and several positive ion peaks were observed. Peaks at m/z = 1432 and 1592 were found adjacent to each other. Calculate the molecular weight of lysozyme. Assume that the charges are due to protonation of the protein and that z differs by one unit for the two peaks.

8.2. A peptide was subjected to degradation from the N-terminus (Edman degradation). The resultant mixture was subjected to MALDI/MS analysis. The following ladder of m/z was observed: 977.2, 1064.3, 1151.4, 1222.6, 1378.9, and 1465.9. Assume that z = 1 in all cases, and derive the sequence of the N-terminal region of the peptide. The identical peptide was enzymatically phosphorylated with protein kinase C. For this peptide, the ladder of m/z was as follows: 977.3, 1064.3, 1231.6, 1302.6, 1458.9, and 1545.9. Explain these results.

8.3. Derive an equation for the time of flight of an ion with a mass/charge ratio of m/z that travels a distance L in the mass spectrometer. Calculate the time of flight for a particle with m/z = 200 amu that was accelerated by 3000 volts over a distance of 30 cm. (Hint: Start with Eq. 8-6 in your derivation. Be careful of the units in making your calculation.)

8.4. Avidin is a glycoprotein found in egg white. It binds an important bio-molecule, biotin, very tightly. Mass spectrometry of avidin using ESI displayed the usual multiple peak spectrum due to multiple charges on a single avidin molecule. Two of the peaks had m/z values of 4002 and 4251. When the avidin was reacted with biotin prior to mass spectrometry, these peaks had m/z values of 4058 and 4312. Determine how many molecules of biotin are bound per molecule of avidin. The molecular weight of biotin is 244.

APPENDIX 1

USEFUL CONSTANTS AND CONVERSION FACTORS

Avogadro's number, N_0	6.0221×10^{23} mole^{-1}
Gas constant, R	8.3144×10^7 erg K^{-1} mole^{-1}
	8.3144 Joule K^{-1} mole^{-1}
	1.9872 calorie K^{-1} mole^{-1}
	0.082057 L atmosphere K^{-1} mole^{-1}
Boltzmann's constant, k_B	1.3806×10^{-23} Joule K^{-1} molecule^{-1}
Planck's constant, h	6.6262×10^{-34} Joule-second
Speed of light, c	2.9979×10^8 m second^{-1}
Standard gravity, g	9.8066 meter second^{-2}
Electronic charge, e	1.6022×10^{-19} Coulomb
Electron mass, m_e	9.1094×10^{-31} kg
Proton mass, m_p	1.6276×10^{-27} kg
Faraday constant, F	9.6485×10^4 Coulomb mole^{-1}

1 calorie = 4.184 Joule
1 Joule = 10^7 erg = 1 volt-Coulomb
1 electron volt = 1.602×10^{-19} Joule

Spectroscopy for the Biological Sciences, by Gordon G. Hammes
Copyright © 2005 John Wiley & Sons, Inc.

APPENDIX 2

STRUCTURES OF THE COMMON AMINO ACIDS AT NEUTRAL pH

Spectroscopy for the Biological Sciences, by Gordon G. Hammes
Copyright © 2005 John Wiley & Sons, Inc.

Aliphatic

Alanine (Ala) (A)	Valine (Val) (V)	Leucine (Leu) (L)	Isoleucine (Ile) (I)

Nonpolar

Glycine (Gly) (G)	Proline (Pro) (P)	Cysteine (Cys) (C)	Methionine (Met) (M)

Aromatic

Histidine (His) (H)	Phenylalanine (Phe) (F)	Tyrosine (Tyr) (Y)	Tryptophan (Trp) (W)

Polar

Asparagine (Asn) (N)	Glutamine (Gln) (Q)	Serine (Ser) (S)	Threonine (Thr) (T)

Charged

Lysine (Lys) (K)	Arginine (Arg) (R)	Aspartate (Asp) (D)	Glutamate (Glu) (E)

COMMON NUCLEIC ACID COMPONENTS

Cytosine
(C)

Guanine
(G)

Adenine
(A)

Thymine
(T)

Uracil
(U)

β-D-ribofuranose

β-D-2-deoxyribofuranose

Spectroscopy for the Biological Sciences, by Gordon G. Hammes
Copyright © 2005 John Wiley & Sons, Inc.

INDEX

Spectroscopy for the Biological Sciences, by Gordon G. Hammes
Copyright © 2005 John Wiley & Sons, Inc.